Springer Theses

Recognizing Outstanding Ph.D. Research

Aims and Scope

The series "Springer Theses" brings together a selection of the very best Ph.D. theses from around the world and across the physical sciences. Nominated and endorsed by two recognized specialists, each published volume has been selected for its scientific excellence and the high impact of its contents for the pertinent field of research. For greater accessibility to non-specialists, the published versions include an extended introduction, as well as a foreword by the student's supervisor explaining the special relevance of the work for the field. As a whole, the series will provide a valuable resource both for newcomers to the research fields described, and for other scientists seeking detailed background information on special questions. Finally, it provides an accredited documentation of the valuable contributions made by today's younger generation of scientists.

Theses are accepted into the series by invited nomination only and must fulfill all of the following criteria

- They must be written in good English.
- The topic should fall within the confines of Chemistry, Physics, Earth Sciences, Engineering and related interdisciplinary fields such as Materials, Nanoscience, Chemical Engineering, Complex Systems and Biophysics.
- The work reported in the thesis must represent a significant scientific advance.
- If the thesis includes previously published material, permission to reproduce this must be gained from the respective copyright holder.
- They must have been examined and passed during the 12 months prior to nomination.
- Each thesis should include a foreword by the supervisor outlining the significance of its content.
- The theses should have a clearly defined structure including an introduction accessible to scientists not expert in that particular field.

More information about this series at http://www.springer.com/series/8790

Scott R. Waitukaitis

Impact-Activated Solidification of Cornstarch and Water Suspensions

Doctoral Thesis accepted by the University of Chicago, USA

Author
Dr. Scott R. Waitukaitis
Department of Physics, James Franck Institute
University of Chicago
Chicago, IL
USA

Supervisor
Prof. Heinrich Jaeger
Department of Physics, James Franck Institute
University of Chicago
Chicago, IL
USA

ISSN 2190-5053 ISSN 2190-5061 (electronic)
ISBN 978-3-319-35927-4 ISBN 978-3-319-09183-9 (eBook)
DOI 10.1007/978-3-319-09183-9

Springer Cham Heidelberg New York Dordrecht London

Softcover reprint of the hardcover 1st edition 2015

Printed on acid-free paper

Springer is part of Springer Science+Business Media (www.springer.com)

To little kids

Supervisor's Foreword

When we think about solids or fluids, we have an intuition about how they should behave: solids are materials that keep their shape and support weight; fluids easily deform and flow under shear to fill the shape of their container. These classifications are reinforced by experience with the world around us and, given many of the materials we are familiar with from daily life, we usually have no reason to rethink these categories.

Not all materials are so easily categorized, however. It turns out that there are large classes that behave neither quite like traditional crystalline solids nor like simple amorphous fluids. One important example are glasses, which respond to applied stress as if they were rigid solids, despite the fact that they are structurally indistinguishable from liquids. Furthermore, in the area of soft condensed matter physics, there has been a realization that more often than not materials defy traditional categorization with complex behavior that lies somewhere between fluid- and solid-like. At the root of this is that soft materials can easily been driven far from equilibrium or often get stuck far from equilibrium when left to relax. Understanding and controlling what happens far-from-equilibrium comprises one of the grand challenges in condensed matter physics today.

Among the conceptually simplest systems with which to probe far-from-equilibrium behavior is a liquid to which solid particles have been added. Adding small glass spheres to oil is one example; another is adding cornstarch particles to (cold) water. As worked out by Einstein over 100 years ago, a small amount of particles increases the liquid's flow resistance, or viscosity, in proportion to the volume fraction occupied by the particles, but otherwise leaves the overall character unchanged. Something altogether new happens once the volume fraction gets sufficiently large that particles start to interact. Now the viscosity of the fluid can become strongly dependent on the strength of the applied shear, and the resulting behavior can either lead to a decrease in flow resistance (shear thinning) or an increase (shear thickening) with forcing.

Shear thickening, in particular, is a striking and highly counter-intuitive phenomenon; in many ways, it is prototypical of the intriguingly complex behavior encountered far from equilibrium. Here is a situation where a particle-laden fluid can be completely liquid-like at rest, but when forced increases its flow resistance to the point where the material can abruptly turn solid-like and even fracture, only to morph back into its liquid-like state once the forcing is removed.

Earlier attempts to model shear thickening, starting in the 1970s within the rheology community, focused on the changes of viscosity due to an order-disorder transition in the particle arrangement. Hydrodynamically induced clustering of particles during shear flow was predicted in the mid-1980s to produce a significant viscosity increase beyond that calculated by Einstein. Such "hydroclusters" were subsequently observed by scattering techniques and very recently, by direct imaging using fast confocal microscopy. However, there is now a consensus that an ordered flow configuration does not have to precede shear thickening and that hydroclusters cannot account for the very large, orders of magnitude increases in viscosity observed experimentally in highly concentrated particle suspensions.

While the hydrocluster scenario starts conceptually from the liquid as the key ingredient that mediates the local interaction among particles, the opposite point of view is embodied in a granular scenario. Here, frictional interactions among particles in direct contact dominate the behavior, and the liquid plays a secondary role. In the granular scenario, the relation between shear and normal stresses becomes nonlocal and boundary conditions such as confinement become key players, determining whether shear thickening can be observed and with what magnitude. Over the last years' work, several groups, including ours at Chicago, have demonstrated that such granular scenario can indeed explain the strong and quite abrupt thickening observed in steady-state shearing experiments.

However, neither scenario can explain the most spectacular aspect of stress response in these systems: impacting the surface of a dense suspension will transform the material and generate enormous normal stresses, at least temporarily preventing the impacting object from sinking and potentially letting it bounce off as if it had hit a solid. In fact, the magnitude of these stresses is large enough to support the weight of a grown person running across a pool filled with the fluid. In prior work, this behavior had typically been interpreted as arising from a viscosity increase due to strong shear. What had been missed is that the required increase would need to be orders of magnitude larger than possible solely via a shear-based mechanism.

The main finding of this thesis is the discovery of a new mechanism that can generate extremely large normal stresses under non-steady-state conditions. Using high-speed imaging and X-ray techniques, systematic experiments revealed that impact triggers a rapidly moving solidification front. As it sweeps through the system, this front converts liquid-like suspension into a quickly growing solid region, which then takes up the impact momentum. Rather than being based on shear, this solidification involves a slight compression of the particle sub-phase inside the suspension, bringing particles into closer contact until they collectively undergo a "jamming transition." This occurs once the network of frictional interparticle contacts prevents relative particle movement and the compressed region turns rigid.

The dynamic jamming process introduced in this thesis for the first time is a direct consequence of far-from-equilibrium behavior in systems of hard particles. The new physics discovered emerged from the process of transformation between fluid and amorphous, glassy solid in response to applied stress. It extends prior work within the community of soft matter researchers, which investigated jamming and the associated onset of rigidity in amorphous solids under quasi-static forcing. The advancing jamming fronts probed here are, in many ways, similar to shock fronts, spreading from the spot of impact at speeds that can easily reach ten times those of the impact itself. While the thesis focuses on cornstarch/water suspensions as a model system, the phenomenon is shown to be quite general, occurring in its most basic form also in two-dimensional systems of dry particles prepared close to, but below, the jamming transition.

As a whole, this work has significantly advanced our understanding of dense suspensions, a prototypical far-from-equilibrium system. In particular, the thesis developed a whole new scenario for stress generation in response to impact. And in doing so, it introduced the new concept of dynamic jamming fronts to both the soft matter physics and rheology communities.

Chicago, USA, June 2014 Prof. Heinrich Jaeger

Abstract

Liquids typically offer little resistance to impacting objects [1–4]. Surprisingly, dense suspensions of liquids mixed with micron-sized particles can provide tremendous impact resistance [5], even though they appear liquid like when left at rest or perturbed lightly. The most well-known example is a dense mixture of cornstarch and water, which can easily provide enough impact resistance to allow a full-grown person to run across its surface. Previous studies have linked this so-called "shear thickening" to experiments carried out under steady state shear and attributed it to hydrodynamic interactions [6, 7] or granular dilation [8, 9]. However, neither of these explanations alone can account for the stress scales required to keep a running person above the free surface. This thesis investigates the mechanism for this impact resistance in dense suspensions. We begin by studying impact directly and watching a rod as it strikes the surface of a dense suspension of cornstarch and water. Using high-speed video and embedded force and acceleration sensing, we show that the rod motion leads to the rapid growth of a solid-like object below the impact site. With X-ray videography to see the dynamics of the suspension interior and laser sheet measurements of the surface profile, we show how this solid drags on the surrounding suspension, creating substantial peripheral flow and leading to the rapid extraction of the impactor's momentum. Suspecting that the solidification below the rod may be related to jamming of the particle sub-phase, we carry out 2D experiments with macroscopic disks to show how uniaxial compression of an initially unjammed system can lead to dynamic jamming fronts. In doing so, we show how these fronts are sensitive to the system's initial packing fraction relative to the point at which it jams and also discover that the widths of these fronts are related to a diverging correlation length. Finally, we take these results back to the suspension, where we perform careful, speed-controlled impact to probe the packing fraction dependence. The solidification we observe in these experiments is consistent with what we see in the 2D experiments, giving further support that the impact response of dense suspensions is caused by dynamic jamming fronts.

References

1. J.W. Glasheen, T.A. McMahon, Vertical water entry of disks at low froude numbers. Phys. Fluids **8**, 2078–2083 (1996)
2. R. Bergmann et al., Controlled impact of a disk on a water surface: cavity dynamics. J. Fluid Mech. **633**, 381–409 (2009)
3. J.R. Royer et al., Formation of granular jets observed by high-speed x-ray radiography. Nat. Phys. **1**(3), 164–167 (2005)
4. J.R. Royer et al., Birth and growth of a granular jet. Phys. Rev. E **78**(1), 011305 (2008)
5. S.R. Waitukaitis, H.M. Jaeger, Impact-activated solidification of dense suspensions via dynamic jamming fronts. Nature **487**(7406), 205–209 (2012)
6. N.J. Wagner, J.F. Brady, Shear thickening in colloidal dispersions. Phys. Today **62**(10), 27–32 (2009)
7. X. Cheng et al., Imaging the microscopic structure of shear thinning and thickening colloidal suspensions. Science **333**(6047), 1276–1279 (2011)
8. A. Fall et al., Shear thickening of cornstarch suspensions as a reentrant jamming transition. Phys. Rev. Lett. **100**, 018301 (2008)
9. E. Brown et al., Generality of shear thickening in dense suspensions. Nat. Mater. **9**(3), 220–224 (2010)

Acknowledgments

My sincerest gratitude goes to my advisor, Heinrich Jaeger, for his support, guidance, and stubborn optimism. Heinrich's enthusiasm spreads to his students and colleagues and makes the Jaeger lab an exciting and wonderful place to become a scientist. Joining Heinrich's lab was probably the best decision I made while at the University of Chicago.

I also want to thank the other faculty, both from the University of Chicago and elsewhere, who played a major role in my development. Wendy Zhang always asked the right questions and forced me to think about my problems more deeply. Sidney Nagel never failed to deliver a good idea when I needed it. Tom Witten always thought outside the box and also gave me the opportunity to go to Chile, where I grew as a person and a scientist. While working in Chile, Nicolas Mújica helped me develop confidence in building experiments and solving problems. I had the pleasure of being a teaching assistant to Henry Frisch several times, where I was able to practice teaching and see good teaching in action. Vincenzo Vitelli and Martin van Hecke showed excitement about my work that inspired me to go into new directions. I'd be remiss if I didn't thank Cheng Chin, with whom I worked my first year at the University of Chicago.

Labmates, lab visitors, and JFI staff have helped me tremendously in accomplishing my work over the past 6 years. I owe John Royer a great deal for his guidance during my first year in Heinrich's lab and for including me in his granular streams project. Helge Grütjen was a great friend and partner during that same year. I owe more than I'd like to admit to Marc Miskin, who is both a good friend and a good scientist, and who has excellent taste in music. The list of these cohorts goes on and on, but hopefully it suffices to acknowledge Carlos Orellana, Pongsakorn Kanjanaboos, Joseph Paulsen, Thomas Caswell, Michelle Driscoll, Sebastian Gonzalez, Athanassi Athanassiadis, Gustavo Castillo, Suomi Ponce-Heredia, Leah Roth, Jake Ellowitz, Xiang Cheng, and Arjun Sharma. I also owe a debt of gratitude to the many postdocs who came and went during my time here, including Eric Brown, Justin Burton, Dustin Kleckner, Martin Lenz, and Nicholas Guttenberg, whose advice and guidance I always found useful. I would never have been able to do the experiments I wanted to do without the expertise of Qiti Guo, Justin Jureller, and Helmut Krebs. No matter how crazy what I wanted to do was, they were not only willing but also extremely capable in helping me do it.

Finally, I thank my parents and family for raising me to work hard and Marilyn for helping me keep it all in perspective.

Contents

Symbols

τ	Shear stress (Pa)
$\dot{\gamma}$	Shear rate (1/s)
η_s	Dynamic viscosity (Pa s)
d	Particle diameter (m)
v_{rel}	Relative velocity between two particles (m/s)
δ	Interstitial spacing between particles (m)
F_l	Lubrication force between two identical spheres (N)
Pe	Péclet number (1)
t_D	Particle diffusion timescale (s)
k_B	Boltzmann constant (J/K)
T	Temperature (K)
g	Acceleration due to gravity (m/s^2)
m_r	Impact rod mass (kg)
r_r	Impact rod radius (m)
ϕ_0	Initial packing fraction of suspension or 2D disk system (1)
H	Suspension fill height (m)
M_{cs}	Total mass of cornstarch used during mixing (kg)
ρ_l	Density of suspending liquid (kg/m^3)
M_l	Total mass of suspending liquid used during mixing (kg)
ρ_{cs}	Specific gravity of cornstarch particles (kg/m^3)
υ_0	Rod or rake impact velocity (m/s)
a_r	Instantaneous rod acceleration (m/s^2)
t	Time after impact (s)
a_{peak}	Peak rod acceleration (m/s^2)
t_{peak}	Time of peak rod acceleration (s)
p_{peak}	Max pressure during rod impact (Pa)
υ_r	Instantaneous rod velocity (m/s)
z_r	Instantaneous rod position below original surface level (m)
ε	Coefficient of restitution (1)
F_r	Instantaneous force on rod (N)

F_b	Force measured on container bottom (N)
z_f	Position of front below suspension surface (m)
h_f	Vertical extend of solidification front in suspension (m)
k	Relative front growth rate (1)
t^*	Time for solidification front to reach container bottom (s)
Δ_t	Rise time for force on container bottom (s)
z_s	Depth of surface depression (m)
r	Radial coordinate from rod center (m)
Δ_z	Vertical displacement of suspension interior (m)
m_a	Added mass (kg)
F_{ext}	External forces during impact (N)
C	Added mass coefficient (1)
ρ_s	Suspension density (kg/m^3)
d_l	Large disk diameter (m)
d_s	Small disk diameter (m)
N_l	Number of large disks (1)
N_s	Number of small disks (1)
x	Axial coordinate (m)
ϕ_J	Packing fraction at jamming (1)
$\upsilon_{x,i}$	x-velocity of ith disk in 2D experiment (m/s)
$\upsilon_{y,i}$	y-velocity of ith disk in 2D experiment (m/s)
ϕ_i	Local packing fraction of ith disk in 2D experiment (1)
V	Coarse-grained velocity (m/s)
ϕ	Coarse-grained packing fraction (1)
x_f	Position of the front (m)
Δ_f	Front width (m)
υ_f	Front velocity (m/s)
U	Rescaled front velocity (1)
x_r	Position of rake (m)
D	Diffusion constant (m^2/s)
δV	Root mean square velocity fluctuations at front center (m/s)
ξ	Longitudinal correlation length (m)
α	Opening angle of cone used in liquid impact experiments ($^\circ$)
F_g	Weight of impact rod (N)
F_{buoy}	Buoyant force on rod (N)
F_f	Frictional force on rod from guiderails (N)
δ_0	Initial interstitial spacing in 1D model suspension (m)
F_i	Force on ith particle in 1D model suspension (N)
υ_i	Velocity of ith particle in 1D model suspension (m)
E	Young's modulus (Pa)

Chapter 1
Introduction

In its simplest definition, a suspension is a mixture of macroscopic, undisolved, hard particles in a liquid. The particles are "macroscopic" in the sense that room temperature thermal energy is not enough to prevent them from sinking. Consequently, if left alone for a long time a suspension will separate into a liquid region sitting atop a dense sedimented region, and in practical applications this has to be avoided by constant mixing or agitation. This feature distinguishes suspensions from colloids, where the particles are small enough (typically less than $\sim 1\,\mu$m) to prevent sedimentation.

Anyone who has ever used a product that reminds to "shake well before using" has likely done so to prevent the sedimentation just mentioned. This is a reflection of the fact that, while it may not seem obvious, suspensions surround us in our daily lives. As a cup of coffee left too long on a desk can attest to, coffee particles eventually begin to accumulate on the cup bottom while the liquid color slowly turns lighter (this is especially true for French press, Turkish, or Greek coffee). A more useful example comes in pharmaceuticals, where dispersing small drug particles in a carrier fluid allows for more efficient delivery and often more pleasant intake. For example, the thick, "powdery" consistency of milk of magnesia or Pepto-Bismol ® is caused by the presence of small, undisolved particles. Other products, such as shampoo, also use the same strategy to deliver their active ingredients. Paints are also typically suspensions, but for very different reasons. Pigment, nothing more than finely ground, colored material, is added not only to change the color, but also to change the "flowability" of the mixture. On a much larger scale, oil extracted from the ground is actually full of dirt particles from which the liquid must be extracted.

Despite the fact that suspensions are found everywhere and made of seemingly simple component parts, they exhibit complex behaviors that are not seen in materials comprised of liquids or particles alone. In particular, the resistance of a suspension to flow can change dramatically depending on how strongly it is driven. This behavior can be both useful and detrimental. It is advantageous in paints, for example, which are designed to flow easily when brushed but then resist flow while left still to dry. In oil extraction, it can be detrimental as over-sedimentation and the resultant

S.R. Waitukaitis, *Impact-Activated Solidification of Cornstarch and Water Suspensions*, Springer Theses, DOI 10.1007/978-3-319-09183-9_1

resistance to flow can cause a transport pipes to clog. Other materials show the opposite behavior and become more resistive when driven strongly. This behavior have inspired researchers to investigate the use of suspensions in applications ranging from ballistic woven fabrics [1] for "liquid body armer" to pipe cloggers [2] for "top kill" operations in overflowing oil wells (such as the BP Deepwater Horizon spill in the Gulf of Mexico in 2010). The study of suspensions is therefore important to be able to control their behavior, taking advantage of it when it is useful and preventing it if it is unwanted.

Beyond their practical importance, suspension behavior is inherently interesting to scientists and laypeople alike because it often defies our intuition. As a short search on YouTube reveals, a normally liquid-like suspension of cornstarch and water can be poured onto a vibrating loudspeaker to produce "cornstarch monsters", stable, fingerlike protrusions that seem to come alive and reach upwards to escape their container (see Fig. 1.1). Perhaps the best known phenomenon, and the topic of this thesis, is the ability of cornstarch and water suspensions to support very large normal stresses to impacting objects and allow, for example, a full-grown person to run across their surface without sinking (see Fig. 1.2). A back of the envelope calculation of how much stress is required to support a running person of average weight reveals that this normally liquid material actually pushes upward with stresses exceeding 50 kPa. Although this crowd pleasing "trick" has been repeated in countless elementary

Fig. 1.1 Cornstarch monsters. Solid, growing protrusions created from an initially liquid-like suspension of cornstarch and water as it is vibrated on a loudspeaker. Photo taken from YouTube (http://www.youtube.com/watch?v=3zoTKXXNQIU)

Fig. 1.2 Running on cornstarch and water. Image of a person running across a suspension of cornstarch and water just as the foot leaves the surface. The initially liquid-like suspension seems to solidify during impact, supporting the weight of the person and preventing them from sinking. Photo credit Benjamin Allen

school science classes and even on television shows such as the Discovery Channel's MythBusters, the physical explanation for such a large normal stress has remained unclear.

Part of the reason for our lack of understanding is that suspensions, despite their deceptively simple components, are inherently complex systems. By definition, they are systems made of liquids and *macroscopic* particles, which means that they are inherently not in thermal equilibrium. Consequently, how a suspension responds to driving is sensitive to small changes in the particle configuration, and this can lead to strong hysteretic effects. Furthermore, suspensions are highly dissipative systems, which gives rise to complicated transient behavior. Most studies are performed in the steady state where the suspension is continually driven until a stable response is observed, even though this pragmatic approach restricts us to understanding only a certain subset of phenomena. Finally, and this is especially true for very dense suspensions such as those studied here, they are extremely sensitive to the particle packing fraction ϕ (defined as the volume of particles divided by the total volume of particles plus liquid). Small spatial heterogeneities can lead to large changes in behavior and that the underlying constitutive equations are not necessarily spatially independent.

Most of what we know about suspension rheology, i.e. the study of how these materials flow, comes from steady state experiments in which the shear stress τ is measured as a function of the applied shear rate $\dot{\gamma}$. The viscosity of the suspension is then defined as

$$\eta_s = \frac{d\tau}{d\dot{\gamma}}. \tag{1.1}$$

(Note we will use η_s to denote the suspension viscosity while η we reserve for the viscosity of the suspending liquid.) For so-called "Newtonian" liquids, such as water, the viscosity is independent of $\dot{\gamma}$. For most suspensions, however, more exotic, "non-Newtonian" behaviors are the norm. In shear thinning suspensions, the viscosity decreases with shear rate. As might be guessed, shear thickening suspensions show the opposite behavior where the viscosity increases as a function of shear rate. These characterizations are useful for quickly generalizing the most recognizable behavior of a given suspension. In practice, however, suspensions are even more complicated still, often exhibiting all three of these types of behavior in different shear rate regimes.

The behavior of dense suspensions like cornstarch and water is often associated with shear thickening [3–12], which comes in two varieties. In *continuous shear thickening*, the suspension packing fraction is generally low and the increase in η_s with $\dot{\gamma}$ is subtle [3–5, 9, 11, 13–19]. On the other hand, in *discontinuous shear thickening* the packing fraction is high and the viscosity can rise by orders of magnitude over very small changes in $\dot{\gamma}$ [7, 8, 10, 12, 20]. Independent models have been developed to explain each of these types of shear thickening. In the case of continuous shear thickening, the most widely accepted explanation frames the viscosity rise as a bulk effect in which increasing shear rate produces groups of particles, or *hydroclusters*, that increase the resistance to flow. For discontinuous shear thickening, recent experimental work has shown that the elevated response can be attributed to frustrated dilation. While these explanations both have success in describing the phenomena seen in the particular, steady state experiments from which they were developed, they cannot explain suspension impact response. In the following sections, we will give more thorough introductions to these models and experiments and show why they fall short.

1.1 Continuous Shear Thickening

In situations where the particle packing fraction is low, the increase in η_s with $\dot{\gamma}$ is subtle and, consequently, the thickening behavior is known as continuous shear thickening. This type of behavior is illustrated qualitatively in Fig. 1.3, where typically one observes a Newtonian regime for low shear rates, a shear-thinning regime for intermediate shear rates, and finally a shear-thickening regime for high shear rates. Two models have been proposed to explain this behavior. The first involves an order-to-disorder transition in the particle arrangement [5, 13–15], while the second involves the formation of particle clusters held together by hydrodynamic forces [9, 11, 16–19]. While the order-to-disorder transition model had traction into the late 1990s, most researchers today tend to focus on the hydrocluster picture. Given this trend, our focus here will be the same.

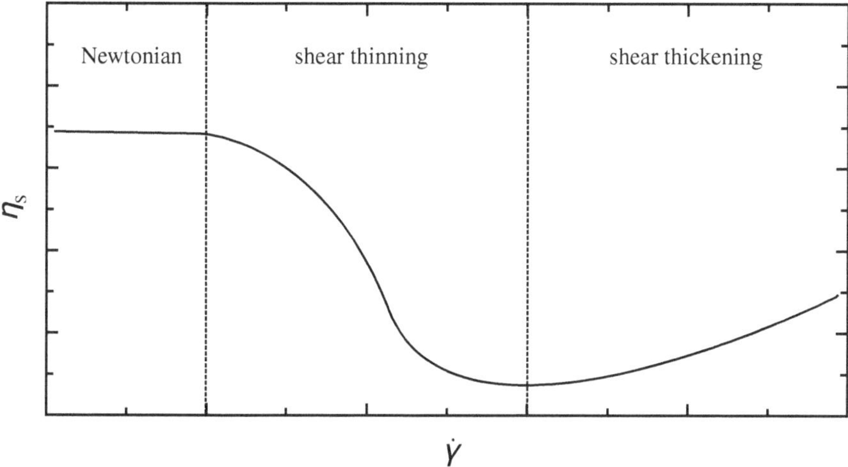

Fig. 1.3 Qualitative graph illustrating the moderate rise in the viscosity η_s versus shear rate $\dot{\gamma}$ for a continuous shear-thickening suspension

As is well known, particles immersed in a liquid will exert a force on each other if they have relative motion. This force arises from the fact that the motion of a single particle causes liquid flow, which in turn influences other particles. In the simplest scenario, one can imagine two spherical particles of equal diameter d immersed in a liquid with viscosity η moving directly toward or away from each other with velocity v_{rel} and with interstitial spacing $\delta \ll d$. In this case, the force between the spheres is exactly solvable [21] and given by

$$F_l = \frac{3\pi\eta d^2 v_{rel}}{8\delta}. \tag{1.2}$$

Thus, when two spheres are in close proximity it becomes increasingly difficult to pull them apart. For more complicated relative motion, e.g. when particles shear past each other, this force can actually lead to stable "orbits" in which the particles are bound together [9].

The transition to continuous shear-thickening behavior has been outlined by Wagner and Brady [9]. At low shear rates, particle motion is dominated by thermal diffusion and the microstructure is essentially random. In this regime, the rheology is Newtonian. As the shear rate is increased, the particles first undergo a layering transition, which tends to reduce the viscosity (i.e. the suspension becomes shear thinning). Eventually, as the shear rate is increased further still, the shearing forces begin to overcome the entropic ones. In this regime bound groups of particles, i.e. hydroclusters, begin to form, and the suspension viscosity rises. This transition can be characterized as a competition between diffusion and advection, and as such the transition to thickening behavior depends on the dimensionless Péclet number. In the context of a shear-driven suspension, the Péclet number compares the shear rate of a flow to the rate at which particles diffuse over the length of one particle diameter, i.e.

$$Pe = \dot{\gamma} t_D = \frac{\eta \dot{\gamma} d^3}{k_B T}, \tag{1.3}$$

where t_D is the diffusive timescale, k_B is Boltzmann's constant, and T is the temperature.

For suspensions with moderate packing fractions, the hydrocluster model has experimental validation. In particular, experiments with small angle neutron scattering have shown an increase in density fluctuations, indicative of groups of particles intermixed with regions that are largely particle free, that is concurrent with thickening behavior [18, 19]. More recently, Cheng et al. used a high-speed, confocal rheometer to image the microstructure of a shear thickening suspension [11]. By directly observing growth in the particle cluster size distribution, they were able to show the increase in viscosity occurred in conjunction with hydrocluster formation.

While these data do support the hydrocluster picture for continuous shear thickening in moderately packed suspensions, they do not show how continuous shear thickening can account for the large stresses generated during surface impact. The biggest issue is that the increase in viscosity is just too small. This has been seen in experiments and simulations [11, 22], but can also be argued quantitatively. Using lubrication theory, the suspension viscosity can be estimated by assuming laminar squeeze flow between layers of particles [23], which results in $\eta_s \approx \eta d/\delta$. However, the continuum model for the liquid will break down when the gap size δ is on the order of 2 molecular layers [24], which means the maximum increase over the suspending liquid viscosity is limited by the ratio of the particle diameter to the molecular diameter. For cornstarch particles in water, this translates into a factor of $\sim 10^4$, or a maximum suspension viscosity of 10 Pa s. If we model a person running on the suspension surface as a sphere (with "foot diameter" 15 cm) moving at 1 m/s through a liquid with this viscosity, the maximum stress we would expect to see is ~800 Pa. As mentioned earlier we can make a conservative, back of the envelope estimate for the stress required to support a person running across a suspension surface by dividing the weight of a typical person (700 N) by the area of a typical foot (200 cm^2), which requires a minimum of ~50 kPa (note this is an underestimate because the upward pressure on the foot must exceed this if the suspension is to decelerate the person).

1.2 Discontinuous Shear Thickening

When the packing fraction is high, experiments show that the increase in η_s with $\dot{\gamma}$ can appear to diverge [7, 8, 10, 12, 20, 25, 26]. Appropriately the effect is called discontinuous shear thickening. Typical behavior is illustrated qualitatively in Fig. 1.4. As with continuous shear thinning, one often encounters a Newtonian regime and perhaps even a shear thinning regime at lower shear rates. At higher shear rates,

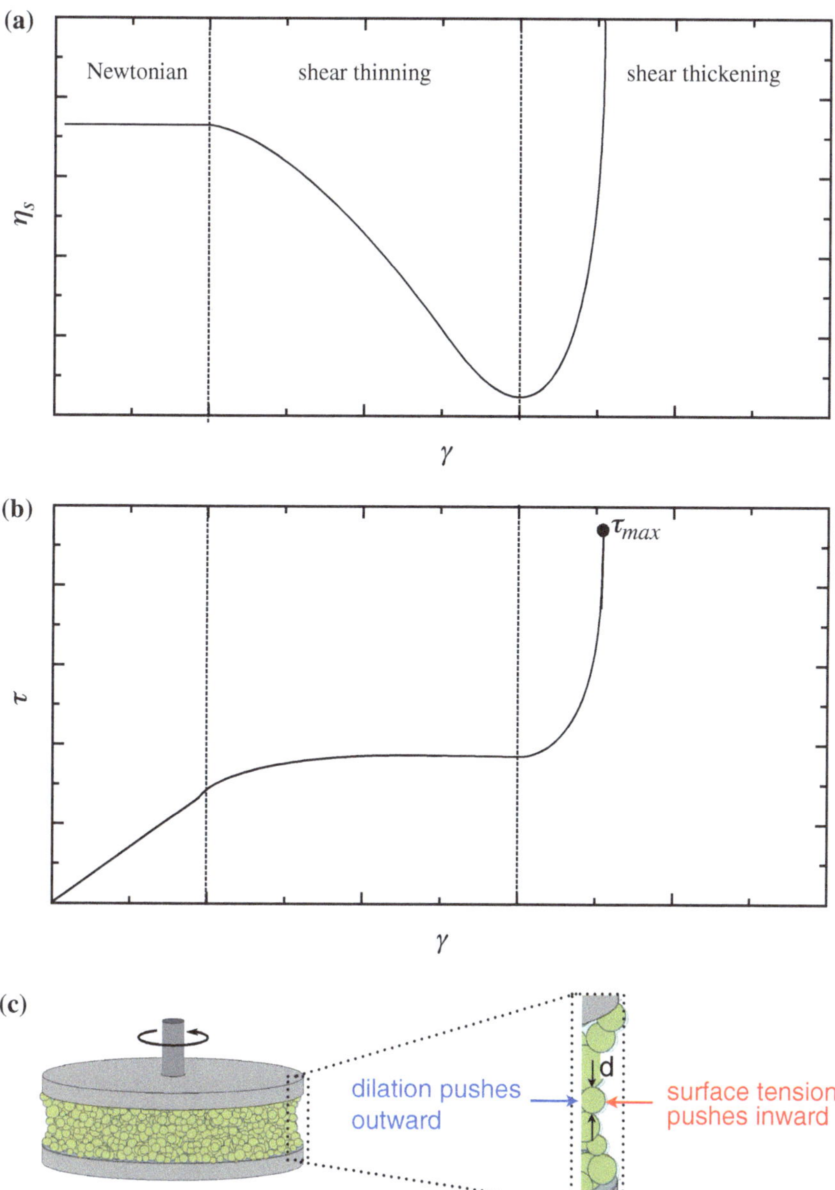

Fig. 1.4 **a** Qualitative graph illustrating the divergent rise in the viscosity η_s versus shear rate $\dot{\gamma}$ for a discontinuous shear thickening suspension. **b** Qualitative graph illustrating the shear stress τ versus shear rate $\dot{\gamma}$ for a discontinuous shear thickening suspension. **c** The stress is limited to be less then $\tau_{max} \approx \sigma/d$, i.e. the stress associated with particles of diameter d poking into the surface of suspending fluid with surface tension σ

however, the viscosity suddenly rises by orders of magnitude. Fall et al. associated this behavior with a re-entrant jamming transition of the particle sub-phase, arguing that the onset of dilatancy during shear could lead to a system wide jamming and a sudden perceived increase in the viscosity [7].

More recently, Brown and coworkers have taken this idea further and showed that the elevated shear stress in this situation is actually caused by the interaction of the particle sub-phase as it dilates and pushes into the system boundaries, specifically the liquid-air interface [8, 10, 20], as illustrated in the cartoon of Fig. 1.4c. As can be argued on dimensional grounds, the pressure provided in this situation scales like

$$\tau_{max} \propto \frac{\sigma}{d}, \tag{1.4}$$

where σ is the surface tension of the suspending liquid. By collecting data throughout a wide range of the shear thickening literature, Brown et al. showed that this idea is on the right track over nearly three orders of magnitude in both surface tension and particle diameter [10].

Another feature of discontinuous shear thickening is that the normal stress and shear stress are seemingly coupled by friction, further supporting a granular picture. Orellana et al. used an electrorheological suspension to show that the normal stress on the plates of a parallel-plate rheometer is proportional to the shear stress [26]. More recently, Brown et al. showed that the exact same behavior could be observed for *dry grains* in a rheometer (confined by solid walls in the absence of a suspending liquid) [26]. Beyond experiments, Seto et al. have recently performed simulations that include both hydrodynamic interactions and frictional contacts [22]. Their results indicate that while hydrodynamic interactions alone can lead to continuous shear thickening, frictional contact among the suspended particles is necessary for producing discontinuous shear thickening.

While the rise in viscosity and stresses encountered in discontinuous shear thickening are much larger than those seen in continuous shear thickening, it still is unable to explain how an object crashing into a suspension surface feels such a large resistive force. Once again, the primary reason for this is that the upper stress limit provided by the free boundary, as discussed above, is just too small. The average particle diameter for cornstarch is $d \sim 15\,\mu m$, while the surface tension of water is $\sigma \approx 0.08\,N/m$. Thus, according to Eq. 1.4 the maximum stress one would expect to see under steady state conditions for a suspension with free boundaries would be on the order of $\sim 5\,kPa$. As we discussed in the previous section, the *minimum* stresses involved in running on the surface are on the order of $\sim 50\,kPa$. Discontinuous shear thickening, like continuous shear thickening, just doesn't produce enough stress to explain impact.

1.3 Compressive Experiments with Shear Thickening Suspensions

While the studies in the previous two sections dealt with suspensions under steady state shear, there are some experiments that have investigated these materials under transient, compressive situations, i.e. conditions more similar to impact. Lim et al. and Jiang et al. [27, 28] studied the transient, compressive response of silica particles suspended in polyethylene glycol with the split Hopkinson bar technique, which essentially consists of sandwiching a thin layer of suspension between two metal rods instrumented with strain gauges and accelerating the rods into each other at high speeds. Although they measured extraordinarily large stresses (up to ~50 MPa), by far the largest reported in the literature, it should be noted that their samples were fully confined, i.e. there was no liquid-air interface. Nonetheless, their results suggested that the suspension, which was liquid-like at rest, suddenly became solid-like during the compression.

More relevant to this thesis, two other recent experiments have studied suspensions under compression with free boundaries. In the first of these experiments, Liu et al. [29] studied the stress transmission below a sphere immersed in a cornstarch and water suspension as it was suddenly pushed at a constant velocity toward the lower boundary of the container holding the suspension. By putting a slab of clay below the sphere, they were able to see how stress was transmitted through the suspension to the opposing boundary via the clay deformation. Perhaps surprisingly, they found that rather than spreading the force on the sphere over a larger area, virtually all the force was concentrated on a spot roughly the size of the sphere directly below it. This, of course, also depended on other factors, such as the speed with which the sphere was pushed and the distance from the clay bottom. Liu et al. interpreted this behavior as arising from a jammed region of suspension transmitting stress directly to the container bottom, but did not address how this jammed region forms.

In the other series of experiments, von Kann et al. studied the dynamics of a sphere as it sank via gravity through a dense suspension of cornstarch and water [30]. As the sphere approached the bottom of the container, they observed oscillations in the sphere velocity in which it would accelerate at nearly g, come to a sudden stop for a few tens of milliseconds, and then accelerate again. In agreement with Liu et al., von Kann et al. interpreted this behavior in the context of jamming. However, they took one step further by thinking about how the jammed region under the sphere develops. As their argument goes, the motion of the sphere toward the wall uniformly increases the packing fraction in the column of suspension below it. Eventually, the particles in this column become jammed and the interaction of the sphere with the boundary through this jammed column causes it to suddenly stop. In the absence of sphere motion, the jammed region has a chance to "melt" with a timescale set by the Darcy flow of liquid back into this region, after which the sphere begins to move again.

1.4 Preview of Our Work

The work presented in this thesis began with experiments designed to understand how an object crashing into a suspension surface is slowed down. In Chap. 2 , we elaborate and expand upon the study we published in Nature [31], where we shoot a rod into a dense suspension of cornstarch and water. Our measurements show that the normal stresses on the face of the impacting rod can be as high as 1 MPa, much larger than the 10 kPa limit in shear thickening. We find that the impact response is not sensitive to the suspending liquid viscosity or the surface tension, further differentiating it from the existing shear thickening models. In agreement with von Kaan et al. [30] and Liu et al. [29], we find evidence that some region of the suspension solidifies in response to compression. However, we show that the primary slowing during impact is associated with the growth of this region, not necessarily an interaction with a boundary.

In Chap. 3, we elaborate and expand on a study we published in Europhysics Letters [32], where we develop a model for the suspension solidification with a macroscopic system of binary-sized disks sitting on a plane and uniaxially compressed with a rake. The compression of the initially unjammed system creates a *dynamic jamming front* that travels ahead of the rake, leaving regions behind it jammed while regions in front of it remain uncompacted. We show how the speed of these fronts depends on the speed of the rake as well as how close the initial disk configuration is to jamming. Using disk conservation in conjunction with an assumed upper bound to the packing fraction at jamming, we account for the front speed as a function of packing fraction with a simple equation. We also find that the width of these jamming fronts seems to diverge as the initial packing density is increased. Interestingly, the growth of the front width is reminiscent of diverging lengthscales seen in other jamming systems.

Finally, in Chap. 4 we take the results from the model system of Chap. 3 and test them in the suspension. By performing speed-controlled impact of a rod into the suspension with an Instron materials testing device, we make more accurate measurements of how fast the suspension solidifies. We show how the solidification rate depends on the suspending liquid viscosity, impact velocity, and packing fraction. In particular, we show that if the product of the viscosity and velocity is too low, then the front growth disappears. Alternatively, if the product of these two quantities is sufficiently large, then the speed of the front depends only on packing fraction. The dependence of the speed of the front on packing fraction in this regime is consistent with the results of Chap. 3, which suggests that the impact response in cornstarch and water suspensions is mediated by dynamic jamming.

References

1. Y.S. Lee, E.D. Wetzel, N.J. Wagner, The ballistic impact characteristics of kevlar® woven fabrics impregnated with a colloidal shear thickening fluid. J. Mater. Sci. **38**(13), 2825–2833 (2003)
2. P. Beiersdorfer, D. Layne, E.W. Magee, J.I. Katz, Viscoelastic suppression of gravity-driven counterflow instability. Phys. Rev. Lett. **106**(5), 058301 (2011)
3. W.H. Boersma, J. Laven, H.N. Stein, Shear thickening (dilatancy) in concentrated dispersions. AIChE J. **36**(3), 321–332 (1990)
4. T.N. Phung, J.F. Brady, G. Bossis, Stokesian dynamics simulation of brownian suspensions. J. Fluid Mech. **313**, 181–207 (1996)
5. R.L. Hoffman, Explanations for the cause of shear thickening in concentrated colloidal suspensions. J. Rheol. **42**, 111–124 (1998)
6. Y.S. Lee, N.J. Wagner, Dynamic properties of shear thickening colloidal suspensions. Rheol. Acta **42**(3), 199–208 (2003)
7. A. Fall, N. Huang, F. Bertrand, G. Ovarlez, D. Bonn, Shear thickening of cornstarch suspensions as a reentrant jamming transition. Phys. Rev. Lett. **100**, 018301 (2008)
8. E. Brown, H. Jaeger, Dynamic jamming point for shear thickening suspensions. Phys. Rev. Lett. **103**(8), 086001 (2009)
9. N.J. Wagner, J.F. Brady, Shear thickening in colloidal dispersions. Phys. Today **62**(10), 27–32 (2009)
10. E. Brown, N.A. Forman, C.S. Orellana, H. Zhang, B.W. Maynor, D.E. Betts, J.M. Desimone, H.M. Jaeger, Generality of shear thickening in dense suspensions. Nat. Mater. **9**(3), 220–224 (2010)
11. X. Cheng, J.H. McCoy, J.N. Israelachvili, I. Cohen, Imaging the microscopic structure of shear thinning and thickening colloidal suspensions. Science **333**(6047), 1276–1279 (2011)
12. A. Fall, F. Bertrand, G. Ovarlez, D. Bonn, Shear thickening of cornstarch suspensions. J. Rheol. **56**, 575–592 (2012)
13. R.L. Hoffman, Discontinuous and dilatant viscosity behavior in concentrated suspensions. ii. theory and experimental tests. J. Colloid. Interf. Sci. **46**(3), 491–506 (1974)
14. E.A. Collins, D.J. Hoffmann, P.L. Soni, Rheology of pvc dispersions. i. effect of particle size and particle size distribution. J. Colloid. Interf. Sci. **71**(1), 21–29 (1979)
15. R.L. Hoffman, Discontinuous and dilatant viscosity behavior in concentrated suspensions iii. necessary conditions for their occurrence in viscometric flows. Adv. Colloid Interfac. **17**(1), 161–184 (1982)
16. D.R. Foss, J.F. Brady, Structure, diffusion and rheology of brownian suspensions by stokesian dynamics simulation. J. Fluid Mech. **407**(1), 167–200 (2000)
17. J.R. Melrose, R.C. Ball, Continuous shear thickening transitions in model concentrated colloids: The role of interparticle forces. J. Rheol. **48**, 937–961 (2004)
18. B.J. Maranzano, N.J. Wagner, Flow-small angle neutron scattering measurements of colloidal dispersion microstructure evolution through the shear thickening transition. J. Chem. Phys. **117**, 10291–10303 (2002)
19. D.P. Kalman, N.J. Wagner, Microstructure of shear-thickening concentrated suspensions determined by flow-usans. Rheol. Acta **48**(8), 897–908 (2009)
20. E. Brown, H.M. Jaeger, The role of dilation and confining stresses in shear thickening of dense suspensions. J. Rheol. **56**(4), 875–924 (2012)
21. L.M. Hocking, The effect of slip on the motion of a sphere close to a wall and of two adjacent spheres. J. Eng. Math. **7**(3), 207–221 (1973)
22. R. Seto, R. Mari, J.F. Morris, M.M. Denn, Discontinuous shear thickening of frictional hard-sphere suspensions. arXiv.org, cond-mat.soft, June 2013
23. N.A. Frankel, A. Acrivos, On the viscosity of a concentrated suspension of solid spheres. Chem. Eng. Sci. **22**(6), 847–853 (1967)
24. J. Van Alsten, S. Granick, Molecular tribometry of ultrathin liquid films. Phys. Rev. Lett. **61**, 2570–2573 (1988)

25. E.E. Bischoff White, M. Chellamuthu, J.P. Rothstein, Extensional rheology of a shear-thickening cornstarch and water suspension. Rheol. Acta **49**(2), 119–129 (2010)
26. C.S. Orellana, J. He, H.M. Jaeger, Electrorheological response of dense strontium titanyl oxalate suspensions. Soft Matter **7**(18), 8023–8029 (2011)
27. A.S. Lim, S.L. Lopatnikov, N.J. Wagner, J.W. Gillespie, Investigating the transient response of a shear thickening fluid using the split hopkinson pressure bar technique. Rheol. Acta **49**(8), 879–890 (2010)
28. W. Jiang, X. Gong, S. Xuan, W. Jiang, F. Ye, X. Li, T. Liu, Stress pulse attenuation in shear thickening fluid. Appl. Phys. Lett. **102**(10), 101901 (2013)
29. B. Liu, M. Shelley, J. Zhang, Focused force transmission through an aqueous suspension of granules. Phys. Rev. Lett. **105**(18), 188301 (2010)
30. S. von Kann, J.H. Snoeijer, D. Lohse, D. van der Meer, Nonmonotonic settling of a sphere in a cornstarch suspension. Phys. Rev. E **84**(6), 060401 (2011)
31. S.R. Waitukaitis, H.M. Jaeger, Impact-activated solidification of dense suspensions via dynamic jamming fronts. Nature **487**(7406), 205–209 (2012)
32. S.R. Waitukaitis, L.K. Roth, V. Vitelli, H.M. Jaeger, Dynamic jamming fronts. EPL (Europhysics Letters) **102**(4), 44001 (2013)

Chapter 2
Freely Accelerating Impact into Cornstarch and Water Suspensions

2.1 Introduction

While the experiments described in the previous chapter primarily focused on rheological measurements of dense suspensions, the focus of this thesis is surface impact. As a number of studies have shown, liquids and granular media typically flow around and provide little resistance to intruding objects [1–11], while suspensions can provide normal stresses that are large enough to support a person running across their surface. As discussed previously, this impact response has been attributed to suspension response under shear, linking it to hydrodynamic interactions [12–17] or a combination of granular dilation and jamming [18–24], but neither of these mechanisms alone can produce enough normal stress to explain impact. In this chapter, we describe a series of experiments designed specifically to study impact into dense suspensions. With techniques ranging from high-speed videography to embedded force sensing and X-ray imaging, we capture the detailed dynamics of the impact process as a metal rod strikes the surface of a dense cornstarch and water suspension. The data reveal that the impactor motion causes the rapid growth of a solid-like region directly below the impact site. These findings are in agreement with von Kann et al. but we go one step further by showing that this is mediated by "solidification fronts" and that no boundaries are necessary for the suspension to provide large normal stresses. Instead, as this solid moves and grows, it pulls on the surrounding suspension creating a quickly growing peripheral flow. Using the concept of added mass, we make a model that relates the sudden extraction of the impactors momentum to the growth of this flowing solid/peripheral region.

2.2 Experimental Setup

In Fig. 2.1a we show a schematic of the experimental apparatus. An aluminum rod (mass $m_r = 0.368\,\text{kg}$, radius $r_r = 0.93\,\text{cm}$) is shot into the surface of a cornstarch and water suspension by gravity or by slingshot. Vertical motion is maintained by

S.R. Waitukaitis, *Impact-Activated Solidification of Cornstarch and Water Suspensions*, Springer Theses, DOI 10.1007/978-3-319-09183-9_2

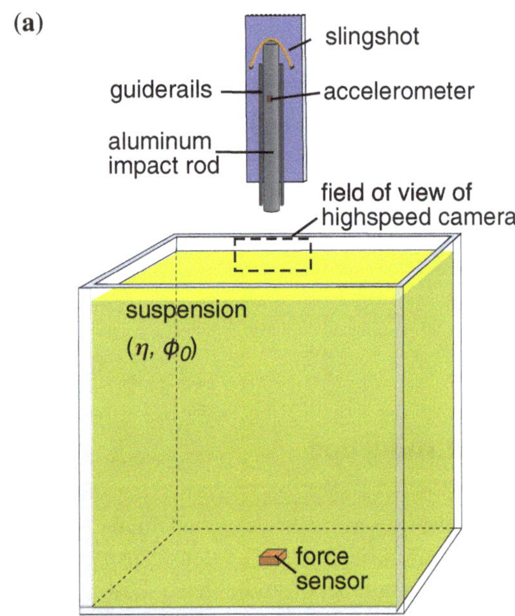

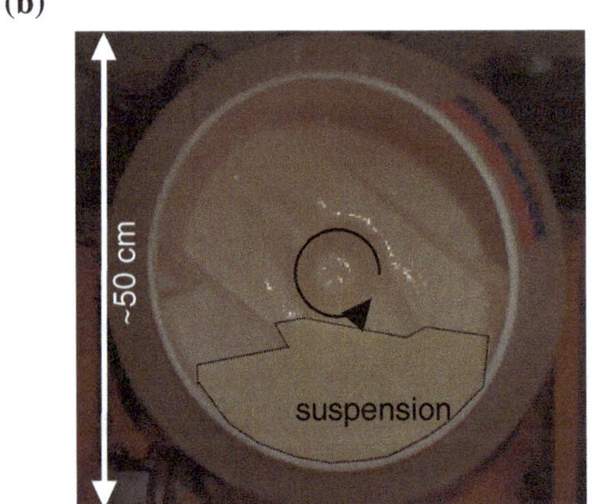

Fig. 2.1 **a** Freely accelerating impact experiment. An aluminum rod ($r_r = 0.93$ cm, $m_r = 0.368$ kg) is accelerated toward the surface of a cornstarch and water suspension (suspending liquid viscosity η, packing fraction ϕ_0, fill height H) via gravity or a slingshot. A high-speed camera focused on the region indicated in the figure tracks the rod to measure the impact velocity v_0. An embedded accelerometer measures the rod's instantaneous acceleration a_r. Directly below the impact site, a force sensor records any stress transmission through the suspension to the container bottom. **b** Preparing ~20 L of cornstarch and water suspension with a cement mixer

gently cradling the rod between stainless steel guide rails. A lightweight, miniature accelerometer (Omega ACC104A) housed inside the rod is connected to a computer via USB data acquisition interface (labVIEW USB-6009) and records the rod's acceleration in real time at a data rate of 24,000 samples per second. As the rod hits the surface, a high speed camera (Phantom v9.1, Vision Research) records video (in the region indicated in the figure) at a typical frame rate of 10,000 frames per second. A laser trigger just above the suspension surface (not shown in the figure) initiates the camera. To measure any stress transmission to the container bottom, we place a high-speed force sensor (DLC101-10, DLC101-50, DLC101-50, or DLC101-500) in a waterproof container directly beneath the rod (like the accelerometer, this sensor is also connected to the USB interface recording at 24,000 samples per second).

The cornstarch and water suspension is characterized by its packing fraction ϕ_0, the suspending liquid viscosity η, the suspending liquid surface tension σ, and the fill height H. The first step in preparing the suspensions involves creating a suspending liquid with the desired viscosity. This is achieved by mixing tap water and glycerin (McMaster Carr 3190K293) with a wire wisk. We then measure the viscosity of this mixture with a capillary viscometer (Cannon Instrument Company, ratings 50–400). After measuring the suspending liquid viscosity, we extract ~25 mL and measure the density with a volumetric measurement from a graduated cylinder and a mass measurement from a scale. (For these experiments, we do not density match the suspending liquid as the sheer volume required makes it prohibitively expensive. We prevent suspension settling by continually relayering the suspension throughout the day with a garden shovel.) We determine the necessary mass M_{cs} of cornstarch by first deciding on the desired packing fraction ϕ_0 and then using the equation

$$M_{cs} = \frac{\rho_{cs}}{\rho_l} M_l \frac{\phi_0}{1 - \phi_0}, \tag{2.1}$$

where M_l is the total mass of the suspending liquid and ρ_{cs} is the specific gravity of the cornstarch particles (i.e. the density of the material itself, not the perceived density of the powder plus air). The value for ρ_{cs} varies throughout the literature from about 1.55 [21] to 1.68 [25]; here, we do not make any measurements ourselves, but instead use a value in the middle of this range $\rho_{cs} = 1.59$ [19]. We measure the total mass of cornstarch with a scale and then slowly added it into the suspending liquid in an industrial cement mixer, as in Fig. 2.1b. We used the cement mixer because the thickening behavior of the suspension makes it extremely difficult to mix "by hand". Adding the cornstarch slowly prevents unwanted clumping and the formation of air bubbles in the suspension. Once all of the cornstarch is added, we let the mixer run for approximately one half hour until the suspension consistency is highly uniform.

2.3 Characterization of Impact

Figure 2.2 shows images before and after the aluminum rod strikes the surface of a deep ($H = 20.5$ cm) cornstarch and water suspension ($\phi_0 = 0.49$, $\eta = 1.0$ cP) with impact velocity v_0 ~1.0 m/s. Rather than penetrating, as would typically happen during impact into liquid or particles alone, the rod pushes the suspension surface downward, creating a rapidly growing depression whose boundary travels away from the impact site. The absence of splashing indicates that the impact is a highly dissipative since none of the incoming kinetic energy is recovered and redirected to ejecta (indeed the collision in a deep container such as this appears almost "inelastic"). Only after the rod has been slowed to a near stop does it begin to actually sink and penetrate into the suspension. We are concerned with the phenomena before this penetration and sinking occurs.

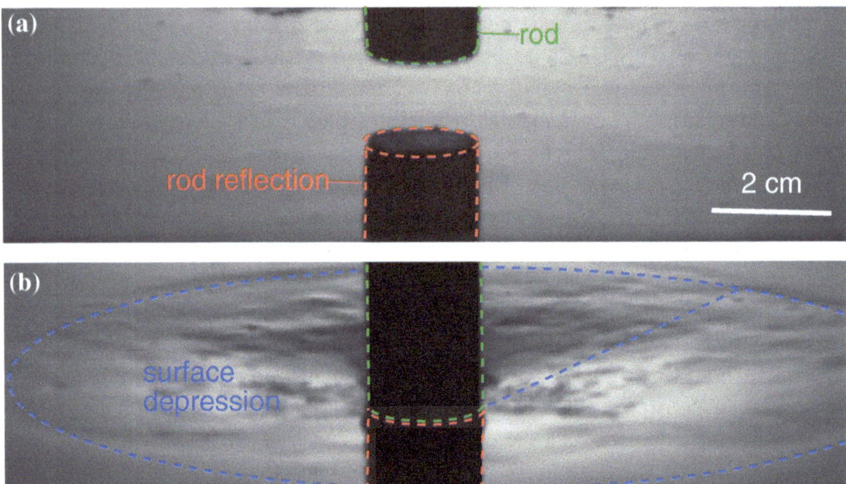

Fig. 2.2 Visual characteristics of suspension impact. Images of an aluminum rod before (**a**) and after (**b**) it strikes the surface of a cornstarch and water suspension ($\phi_0 = 0.49$, $\eta = 1.0$ cP) at $v_0 \sim 1.0$ m/s. Rather than penetrating and creating a splash, the rod seems to push the surface down, creating a large depression that travels radially outward from the impact site. The pockmarked appearance may be a signature of dilation causing particles to poke into the liquid-air interface

Given the complex force laws involved during impact into liquids or particles alone [1–11], one might expect the force law for an object impacting into a suspension to be similarly complex. Surprisingly, this is not the case. Figure 2.3 shows the rod's instantaneous acceleration a_r as measured by the embedded accelerometer plotted against time t (upward acceleration is defined as positive). As the plot shows, the rod's acceleration starts out at minus g before impact ($t < 0$). Just after impact, the acceleration steadily grows to some peak value a_{peak} at time t_{peak} and then slowly decays to some near-zero or slightly negative value. The existence of peaks in the a_r versus t curves indicates that the force law responsible for slowing the rod is a competition between both time-increasing and time-decreasing contributions, and we can use the behavior of these peaks to characterize each impact. As might be guessed from experience, a primary factor that affects the peaking behavior is the impact velocity v_0; higher impact velocities lead to larger peaks that occur at earlier times, as shown in Fig. 2.4a. It is also worth noting here that high velocity impacts can lead to incredibly large pressures on the rod face [$P_{peak} = m_r a_{peak}/(\pi r_r^2)$], up to as much as 1 MPa and thus far exceeding the maximum stress (~5 kPa) encountered in steady state shear experiments [24]. For the highest impact velocities (above about 3.0 m/s), the rod begins to penetrate into the suspension. This transition is especially apparent in the plot of t_{peak} versus v_0 (Fig. 2.4b). Before the transition, t_{peak} decreases with v_0, but beyond it actually begins to increase. (Although we do not at present understand the physics of this transition to penetration, we provide some experimental results concerning it in Appendix A and show that it is sensitive to r_r.)

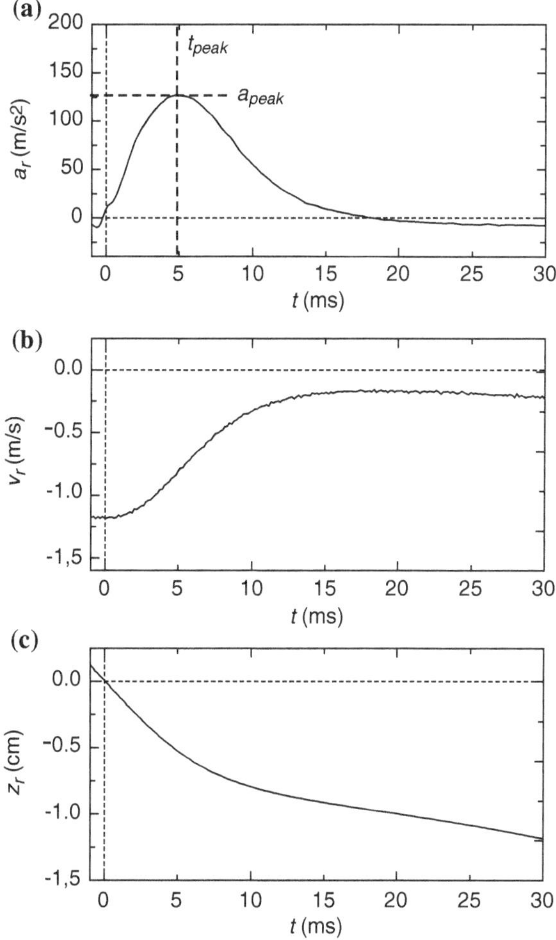

Fig. 2.3 Impact dynamics. **a** Acceleration a_r versus time t for the aluminum rod striking the surface of a cornstarch and water suspension ($\phi_0 = 0.49$, $\eta = 1.0\,\text{cP}$) at impact velocity $v_0 = 1.18\,\text{m/s}$. The impact produces a peak of value a_{peak} occurring at time t_{peak}, indicating an underlying competition in the force law. **b** Impactor velocity v_r versus t, showing impact nearly brings rod to a complete stop before gravity reaccelerates it downwards and it slowly sinks into suspension (not shown). **c** Rod position z_r versus t

Also in accordance with experience, the impact response is highly sensitive to the particle packing fraction ϕ_0, with more densely packed suspensions leading to higher values of a_{peak} and smaller values of t_{peak} (at a given impact velocity v_0, also shown in Fig. 2.4). In densely packed suspensions, an increase of just a few percent in ϕ_0 can cause the peak accelerations to double or triple. This creates practical limitations for conducting experiments. Below $\phi_0 \sim 0.46$, the effect becomes so small that it is difficult to detect. At the other extreme, suspensions with packing

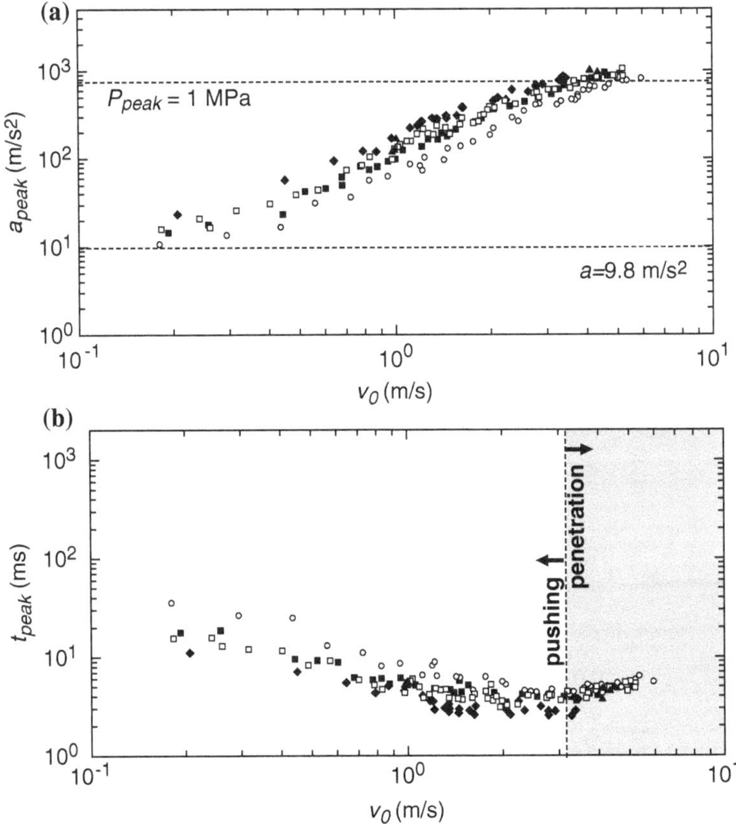

Fig. 2.4 Parameters affecting peaking behavior. **a** Peak accelerations a_{peak} versus impact velocity v_0 for experiments with following parameters: $\eta = 1.0$ cP and $\phi_0 = 0.46$ (*open circles*), $\phi_0 = 0.49$ (*solid squares*), $\phi_0 = 0.52$ (*solid diamonds*), $\eta = 12.4$ cP and $\phi_0 = 0.49$ (*open squares*), and $\eta = 1.0$ cP and $\phi_0 = 0.49$ with a water layer approximately 1 cm deep on the suspension surface (*solid triangles*). **b** Time to peak acceleration t_{peak} versus impact velocity v_0 with same symbols as in (**a**). The vertical *dashed line* indicates the crossover region to the right of which the rod begins to pierce the suspension surface

fractions above ~0.52 are so sensitive to perturbations that their relaxation time for reaching a liquid state can take hours. Consequently, our work is restricted to the regime $0.46 < \phi_0 < 0.52$.

While the peaking behavior is strongly sensitive to the impact velocity and packing fraction, it is surprisingly insensitive to the properties of the liquid. Changing the viscosity by more than a factor of 10 has no observable effect on the peaks (although it does slow the post-impact sinking behavior). Furthermore, completely removing the effects of surface tension by adding a layer of water (~1 cm deep) to the top of the suspension shows that the response is not associated with particles poking out of the liquid-air interface. As mentioned in Chap. 1, the viscosity of a suspension that exhibits continuous shear thickening does scale with the suspending liquid viscosity [26–28], while the stresses of a discontinuous shear thickening suspension is limited by surface tension effects [24]. The irrelevance of both viscosity and surface tension reemphasizes that impact response cannot be explained by existing shear thickening models.

2.4 Boundary Effects

To see what role, if any, is played by the opposing boundary during impact, we changed the suspension height H and looked for corresponding changes in the a_r versus t curves, as in Fig. 2.5. For the deepest suspension, $t_{peak} \sim 10$ ms, but a second, weaker peak is just visible near $t \sim 75$ ms. Lowering H causes this secondary peak to intensify and move to earlier times. Interestingly, however, the first peak remains largely unchanged. For the smallest H, a third peak emerges as a consequence of the impulse on the rod being so large that it bounces upward off of the surface, freely accelerates downward at minus g for a few milliseconds, and then hits the surface again (as indicated in the lowest panel of Fig. 2.5). For the lowest two values of H, the response becomes dominated by this second peak and the character of impact changes completely. Rather than appearing inelastic, the impact can actually be quite elastic, with as much as 25 % of the impactor's initial speed recovered in the recoil (see Fig. 2.6). Additionally, we see that the coefficient of restitution ϵ, defined as the recoil velocity divided by the impact velocity, is largely independent of impact velocity until $v_0 \sim 3.0$ m/s, i.e. near the transition to the penetration regime.

One might guess that the second peak, which occurs while the rod is still in contact with the suspension surface, might arise from the transmission and reflection of waves to the opposing boundary. If this were the case, then one would expect to see a strong, peak-like signal on the force sensor at the container bottom at half the time of the second peak in the accelerometer. Surprisingly, as the bottom panel of Fig. 2.5 shows, the peak on the force sensor occurs at the same time as the second peak. This is actually a signature of the solidification suspected by Liu et al. [29] and von Kann et al. [30], but these measurements have several key, new implications: first, the primary response (i.e. the original acceleration peak) is not the result of stress transmission to the boundary; second, the suspension does indeed solidify, but the solidification process requires a finite amount of time to propagate through the suspension; third, once solidification reaches the bottom boundary, forces propagate with little delay through the solid-like region back towards the impactor; and fourth, this solid can bear stress and store energy, allowing, for example, the bounce of the impactor.

These implications are more fully appreciated in Fig. 2.7, which shows the force on the rod ($F_r = m_r a_r$) and the force on the container bottom for system parameters ($H = 11.5$ cm, $v_0 = 2.0$ m/s, $\phi_0 = 0.49$ and $\eta = 1.0$ cP) that prevent the rod from bouncing and separating from the suspension surface (like the bouncing in Fig. 2.5, which removes the solid-coupling between the rod and container bottom). A slow

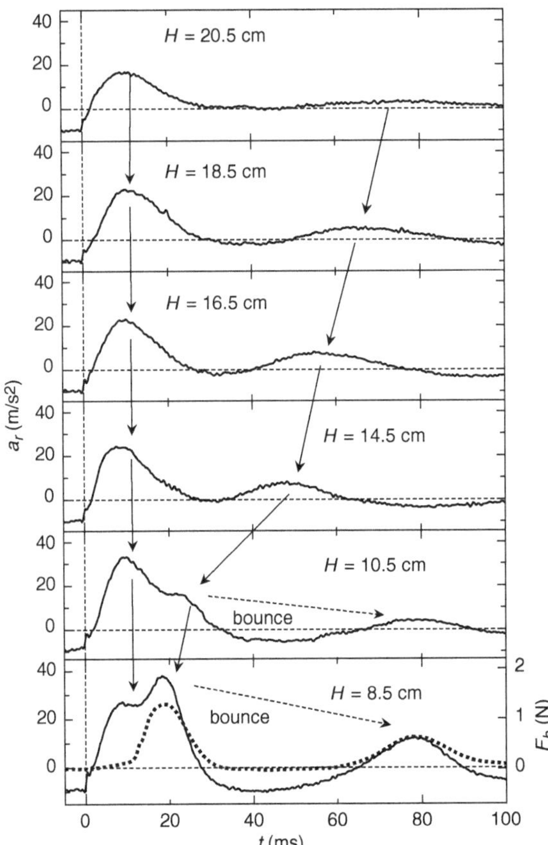

Fig. 2.5 Effect of lower boundary. Rod acceleration a_r versus time t for impact with $\eta = 1.0\,\text{cP}$, $\phi_0 = 0.49$, $v_0 = 0.49 \pm 0.04\,\text{m/s}$ and suspension fill heights $H = 20.5$, 18.5, 16.5, 14.5, 10.5, and 8.5 cm (*top* to *bottom*, as indicated). The *dashed line* in the bottom panel is the force on the sensor at the container bottom F_b

initial buildup of the force measured on the container bottom F_b is followed by an abrupt jump (over $\sim$1.5 ms) to its maximum value of $\sim$7 N at $t \sim 7.5\,\text{ms}$. Before this, F_b and F_r show no correspondence. After this, however, it is clear that the spur on the latter part of F_r has the same shape F_b. This further indicates that the transmission of stress between the rod at the top of the suspension and the force sensor at the bottom is solid-like. What is more, these data suggest that the solid-like region is concentrated in a column almost directly below the rod, in agreement with the clay-witness experiments of Liu et al. [31]. Noting that the area of the force sensor is $\sim 1.13\,\text{cm}^2$ and, assuming the pressure on the bottom is roughly constant, we estimate that the total force on the rod is recovered over an area $\sim 10\,\text{cm}^2$. This is much smaller than the full area of the container bottom ($900\,\text{cm}^2$), and if we imagine the stress propagates through the suspension in a cone this corresponds to an angle

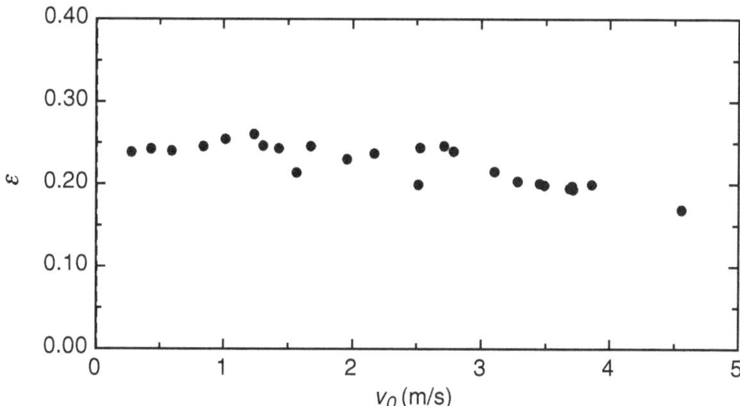

Fig. 2.6 Restitution coefficient. Coefficient of restitution $\epsilon \equiv v_{recoil}/v_0$ (where v_{recoil} is the velocity with which the rod bounces upward off of the surface) for a suspension with $\phi_0 = 0.49$, $\eta = 1.0$ cP, and $H = 8$ cm. The existence of the bounce indicates that the region of suspension below the rod stores elastic energy just like a solid

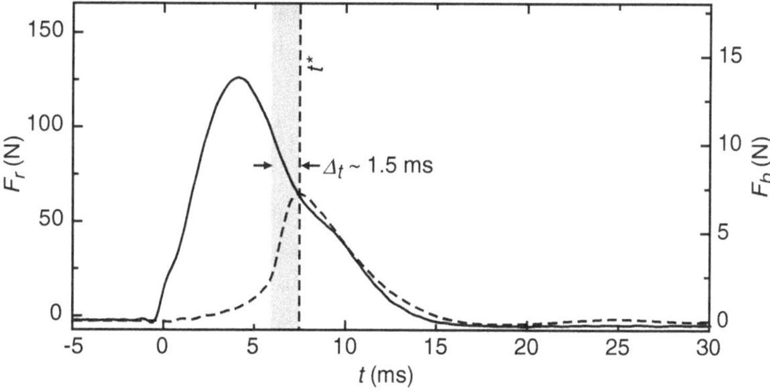

Fig. 2.7 Details of stress transmission to container bottom. Force on rod F_r versus time t (*solid curve, left axis*) and simultaneously measured force on container bottom directly below rod F_b versus t (*dashed curve, right axis*) for suspension with $H = 11.5$ cm, $\phi_0 = 0.49$, $\eta = 1.0$ cP and impact velocity $v_0 = 2.0$ m/s. The time of the peak in F_b (or equivalently, time of 2nd peak in F_r) can be interpreted as time required for solid-like growth to reach bottom. The rise time to the peak (especially pronounced for these impact parameters) can be used to show that the width of the solidification front in this realization is approximately $v_0 \Delta_t \sim 4$ mm

of about 10° (though this may underestimate the cone angle given that the pressure is presumably highest directly below the rod).

These data suggest that the timing of the 2nd peak in the rod acceleration (or equivalently the timing of the first peak on the force sensor below) can be interpreted as the time t^* required for the front of a solid-like column to grow from the suspension surface to the bottom boundary. By measuring t^* at several different H (as we did in Fig. 2.5), we can plot the trajectory of the solidification front as it

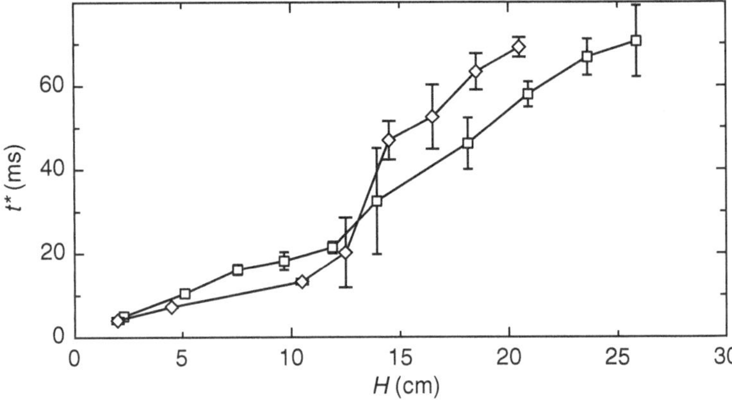

Fig. 2.8 Trajectory of solidification front. Time of the second peak of the rod's acceleration t^* versus suspension fill height H for impact velocities $v_0 = 0.49 \pm 0.04$ m/s (*squares*) and 0.9 ± 0.1 m/s (*diamonds*)

develops, as in Fig. 2.8. Close inspection of these data reveals two points. First, higher impact velocities produce fronts that travel with higher initial speeds. Second, these trajectories have the same qualitative features of the rod trajectories, i.e. two straight line regions connected by a soft bend (see for example Fig. 2.3c). What's more, the timing of the bends in the front trajectories (~10–15 ms) is very close to the timing in the trajectories of the rod itself.

The resemblance of the rod trajectory in Fig. 2.3c and the front trajectory in Fig. 2.8 suggests that the growth of the solid may be related to the displacement of the rod. To test this directly, we plot the size of the solidified region h_f at t^* versus the distance travelled by the rod at the same instant z_r^* in Fig. 2.9 for the same data as in Fig. 2.8. (Note the position of the front below the original surface at t^* is $z_f = -H$. More often, we will refer to the vertical extent of the front $h_f = H - z_r^*$.) Doing so collapses the data for the two different impact speeds onto what is nearly a straight line of given by $h_f = kz_r^*$ with $k \approx 12.2$. We define the proportionality constant k as the *relative front growth rate* of the suspension. Throughout the rest of this thesis, we will refer to k often as it turns out to be an important parameter for characterizing the suspension behavior.

Going back to Fig. 2.7, we can use the rise time in the force measured on the container bottom to estimate the width of the solidification front. As the figure shows, $\Delta_t \sim 1.5$ ms. Given that the front crashes into the bottom with the speed of the rod, its width is approximately $v_r \Delta_t \sim 4$ mm. This shows that, relative to the size of the solidified column or even the size of the rod, the solidification front can be extremely well-defined. (As we will show in this chapter and Appendix B, the front width may depend on both the particle packing fraction and the viscosity of the suspending liquid, though we did not experimentally probe these dependencies in great detail.)

These behaviors are reminiscent of shocks in granular systems above jamming [32, 33] or solidification fronts in supercooled glass-forming liquids [34–37].

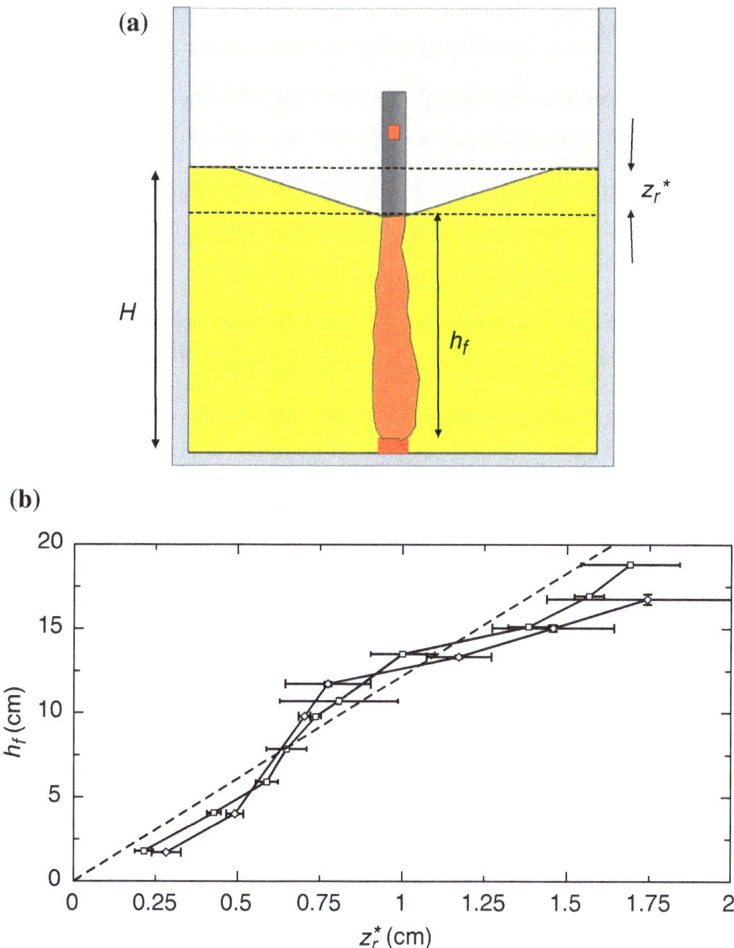

Fig. 2.9 Solid growth versus rod motion. **a** Cartoon of solid growth below impact site. **b** Vertical extent of solidified region h_f at instant front hits container bottom plotted against distance below surface travelled by rod at same instant z_r^* for impact velocities $v_0 = 0.49°$ æ 0.04 m/s (*squares*) and 0.9° æ 0.1 m/s (*diamonds*). Fit is of the form $h_f = kz_r^*$ with $k = 12.2$

However, with granular shocks above jamming the front propagates through an already-jammed medium and its speed is governed by elastic energy stored in particles [32, 33]. Although supercooled liquids, like the system here, are initially unjammed, their solidification fronts propagate at a constant, thermodynamically favored speed [37]. The data in Fig. 2.9 show that the fronts here seem to work like a "snowplow", where the extent of moving snow (suspension) is proportional to how far the shovel (rod) has pushed. As will be shown in this chapter and Chap. 3, this behavior can be tied to jamming of the initially unjammed particle sub-phase as it is compressed by the impacting rod.

2.5 Surface Dynamics

The previous section describes what happens directly below the impact site, but it is not immediately apparent how this relates to the features seen on the suspension surface (e.g. Fig. 2.2). Qualitatively, one can argue that the solidified region must cause this surrounding suspension to flow because the lubrication forces described by Eq. 1.2 act as a kind of "glue" between the closely packed particles. In order measure the depression dynamics, we shine a laser sheet across the suspension surface in the field of view of the high speed camera, as in Fig. 2.10. This allows us to make space-time plots of the depth of the surface depression z_s versus radial coordinate r and time t. A typical result is shown in Fig. 2.11. This plot shows more clearly how, for the region to the left of the dashed line in Fig. 2.4, the rod pushes the suspension surface down rather than penetrating into it (note the continuous color spectra across the dashed line indicating the rod/suspension boundary). The plot also shows how regions outside of the depression (i.e. beyond the $z_s = 0\,\text{mm}$ contour in the blue area) actually swell upward slightly to conserve volume. Interestingly, the trajectory of the $z_s = 0\,\text{mm}$ contour, like the solid front below the rod, is approximately proportional to the total distance travelled by the rod (and with a proportionality factor very close to the relative front growth rate, $k = 12.2$, as indicated by red dashed line in the figure).

2.6 Displacements of Suspension Interior

In order to see what happens inside the suspension, we used a C-arm dental X-ray (Orthoscan High Definition Mini C-Arm, Model 1000-0004) to take video (30 frames per second) of a tracer particles in a vertical plane of the suspension interior directly below the impact site, as shown in Fig. 2.12a. (For technical reasons, we had to make a few changes from the setup in Sects. 2.2–2.4. These changes slightly altered the rod dynamics, as discussed in Appendix B. Even so, the salient features of the impact process remained the same.) The tracer particles consisted of small metal objects (e.g. metal spheres, nuts, screws, bolts, nails, and washers) that slowly sank into (and then out of) the field of view of the X-ray apparatus. While the tracers were in the field of view, we released the rod in free-fall from a fixed height allowing it to impact into the suspension ($v_0 \approx 0.5\,\text{m/s}$) while simultaneously capturing X-ray video. Given the frame rate was limited to 30 frames per second and the typical impact only lasts $\sim$20 milliseconds, these videos give a "before and after" look into the suspension. As the field of view of the X-ray videos was limited to $\sim 5 \times 5\,\text{cm}^2$, we took videos in four separate regions (the boundaries of which can be seen in Fig. 2.12b) and stitched them together.

We used particle imaging velocimetry (PIV) to determine the displacement field of the suspension interior for each video (code in Mathematica written by Justin Burton). Sinking made it difficult to load the tracer particles uniformly, and it was necessary to

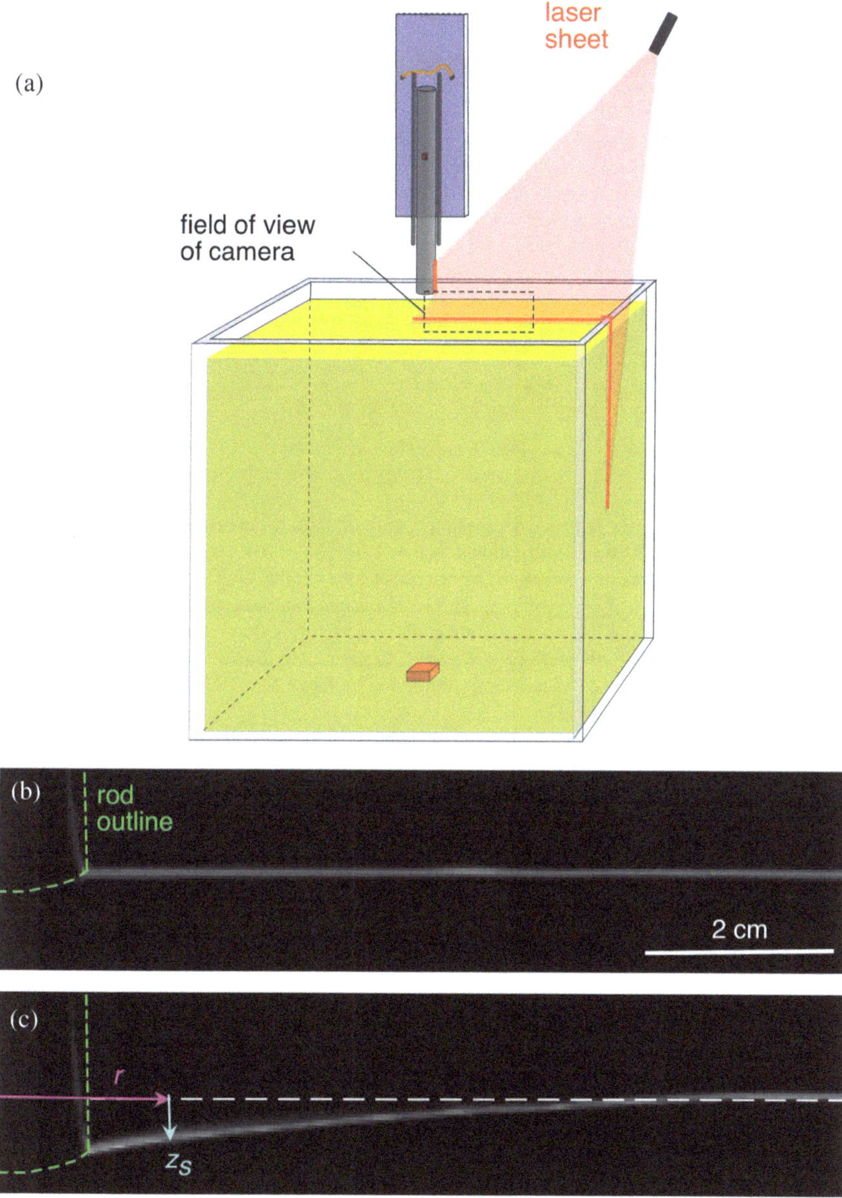

Fig. 2.10 Laser sheet measurements of surface depression. **a** A laser sheet is used to generate a bright line on the suspension surface in the field of view of the camera. **b** Image of laser line on surface just before impact. Rod is outlined in *green*. **c** Image of laser line on surface about 15 ms after impact

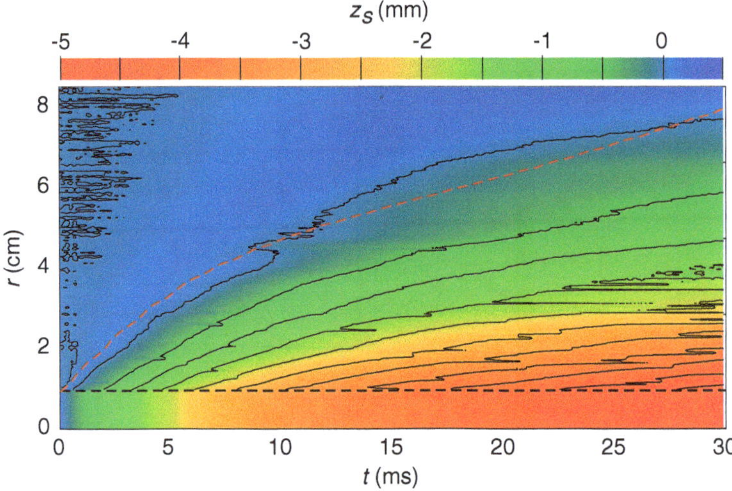

Fig. 2.11 Space-time plot of surface depression. Depth of surface depression z_s *(color axis)* versus radial coordinate r and time t (with parameters $\eta = 1.0\,\text{cP}$, $\phi_0 = 0.49$, and $v_0 = 0.49 \pm 0.01$ m/s). The *blue* part of the figure corresponds to regions outside the conical depression (not there is slight upswell in this region to conserve volume). The *black dashed line* indicates the boundary between the rod and suspension, and the smooth color gradient across this line indicates the rod does not penetrate but instead pushes the surface downward. Contours are drawn for every 1 mm. The *red dashed line* corresponds to the rod trajectory multiplied by factor $k = 12.2$, the same factor found for the solid growth below the rod

take several videos in each field of view, ignore the PIV data corresponding to regions that lacked particles, and then average the results from different videos together to fill in the gaps. A final displacement field is shown in Fig. 2.13. The first striking feature of these data is the large region of suspension that moves downward, extending approximately 6 cm below the rod and 5 cm to the side (red to green in the figure). What makes this especially remarkable is that all of this movement is a result of the rod moving a mere 5 mm below the original surface level. To the side of this downward moving region, the PIV data make it clear that the suspension flows upward. As mentioned in Sect. 2.4, where the same upward motion was seen with the laser sheet measurements (Fig. 2.11), this must occur because the suspension as a whole is incompressible and the vacated volume of the depression has to be compensated for by an upswell on the periphery. These observations give a quick, qualitative answer to how the rod is slowed down during impact. In brief, even a very small amount of rod motion creates a vastly larger region of flow in the suspension. The mechanism for the slowing of the impactor is the transfer of momentum to this growing, moving region.

The PIV data also provide a second opportunity to quantitatively confirm the relationship between the growth of the solid front relative to the displacement of the rod. We start with the following simple assumption: if a segment of suspension has been solidified, it moves rigidly with the rod, whereas if it has not been solidified then

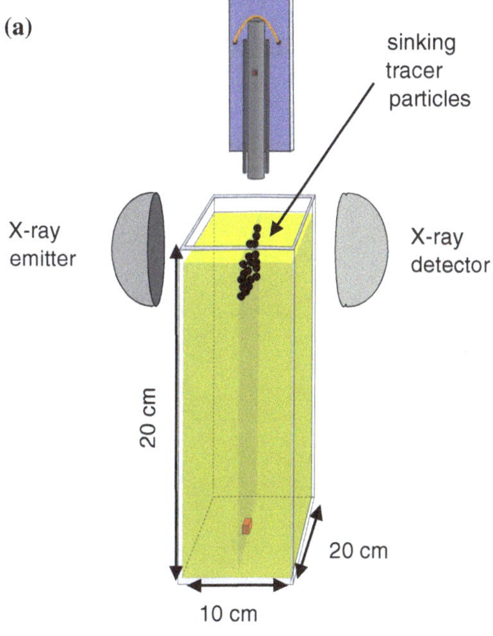

Fig. 2.12 X-ray measurements of interior dynamics. **a** X-ray emitter shines through suspension toward detector. The plane of the suspension directly below the impacting rod is laden with metal particles to act as tracers in the X-ray video. **b** X-ray image of tracer particles in the suspension interior just after rod strikes the surface at $v_0 \approx 0.5\,\mathrm{m/s}$. Note that the container extends equally to the *right* and *left* of the rod, but imaging was performed primarily to the *right* side, hence the asymmetry in the figure

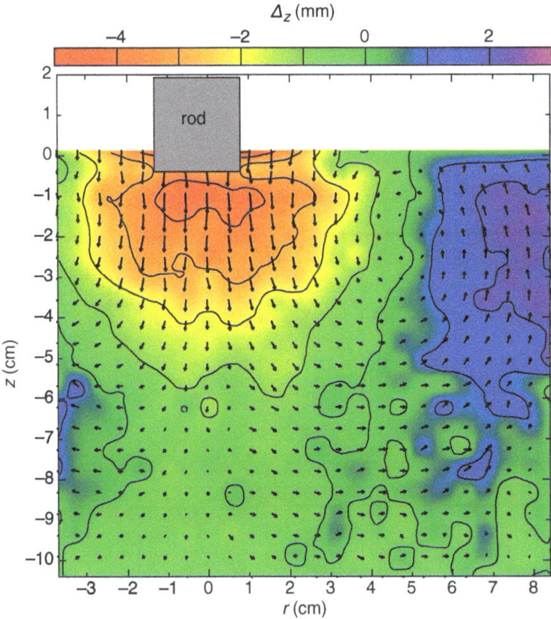

Fig. 2.13 Suspension interior displacement field. Displacement field of suspension in a plane directly below impact site calculated via particle imaging velocimetry (PIV) of X-ray images taken before and after impact. The large downward moving region (*red* to *green* in the figure), which extends nearly 6 cm below and away from the rod, develops after the rod itself moves only ~5 mm. Outside of this downward moving region, the suspension moves upward to conserve volume globally

it does not move all. If the rod moves a total distance $|z_r|$ beyond the original surface between two X-ray images, then the edge of the front will reach a depth $|z_r|(k+1)$ below the surface, so beyond this depth all displacements Δ_z should be zero. Above this depth, a segment of the solidified column at depth $|z|$ will move however far the rod moved after it was picked up, i.e.

$$|\Delta_z| = |z_r| - \frac{|z|}{k+1}. \tag{2.2}$$

In Fig. 2.14 we plot the vertical displacements $|\Delta_z|$ of the suspension below the rod as a function of $|z|$. The data have the qualitative shape predicted by Eq. 2.2, starting out by decreasing linearly and then coming to (nearly) zero displacement. Fitting the linear region to Eq. 2.2, we find $|z_r| = 5.0 \pm 0.2$ mm and $k = 13.1 \pm 0.9$, very close to the value found from varying the suspension height H (Fig. 2.9).

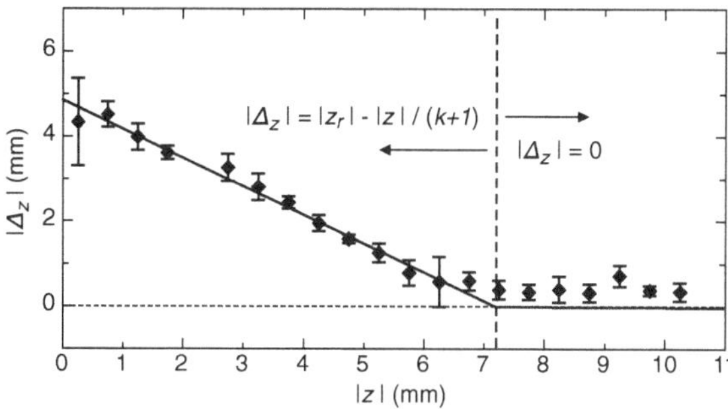

Fig. 2.14 Vertical displacements of suspension directly below impact site. Experimental data (*diamonds*) and model prediction (*black line*). As discussed in the text, the "snowplow" model for solidification predicts that the displacements should be zero beyond the distance $|z| = (k+1)|z_r|$ and should decrease like $|\Delta_z| = |z_r| - |z|/(k+1)$ before this distance. Fitting to this proposed form gives the relative front growth rate $k = 13.1 \pm 0.9$, close to the value found in Sect. 2.2

2.7 Added Mass Model for Impact

The results of the previous three sections paint a picture in which the seed of the suspension response to impact is the dynamic growth of the solid below the impact site. As this solid grows and is forced to move with the rod, it causes flow in the surrounding, still liquid-like suspension. The interplay between this growing region of moving suspension and the slowing of the rod is the competition mechanism responsible for the observed peaks in the rod deceleration. We can capture the essence of this behavior using the concept of added mass, as is frequently done for surface impact in regular liquids [2, 4, 38]. The key idea is to think of the impact as an inelastic collision between the rod and a growing mass, m_a. The rod dynamics are captured by force balance:

$$(m_r + m_a)a_r = \frac{dm_a}{dt}v_r + F_{ext}. \tag{2.3}$$

where F_{ext} accounts for other forces not associated with momentum transfer to the added mass, e.g. the force of gravity on the rod $m_r g$ and the buoyant force from the displaced liquid in the depression (from Fig. 2.5, this is $\sim 1/3\pi g(r_r + k|z_r|)^2|z_r|$). With normal liquids, m_a is typically limited by the density of the liquid and the size of the impactor, for example, $m_a < C(4/3)\pi\rho_l(r_r)^3$ for the impact of a disk of radius r_r into a liquid of density ρ_l [2]. The factor C is the "added mass coefficient" and accounts for the fact that the liquid does not actually move like a solid object (consequently, C is typically less than 1). The suspension is capable of responding so dramatically because the solidification below the rod leads to rapid, effectively unlimited growth of m_a. We can estimate its size from Figs. 2.9 and 2.11, which

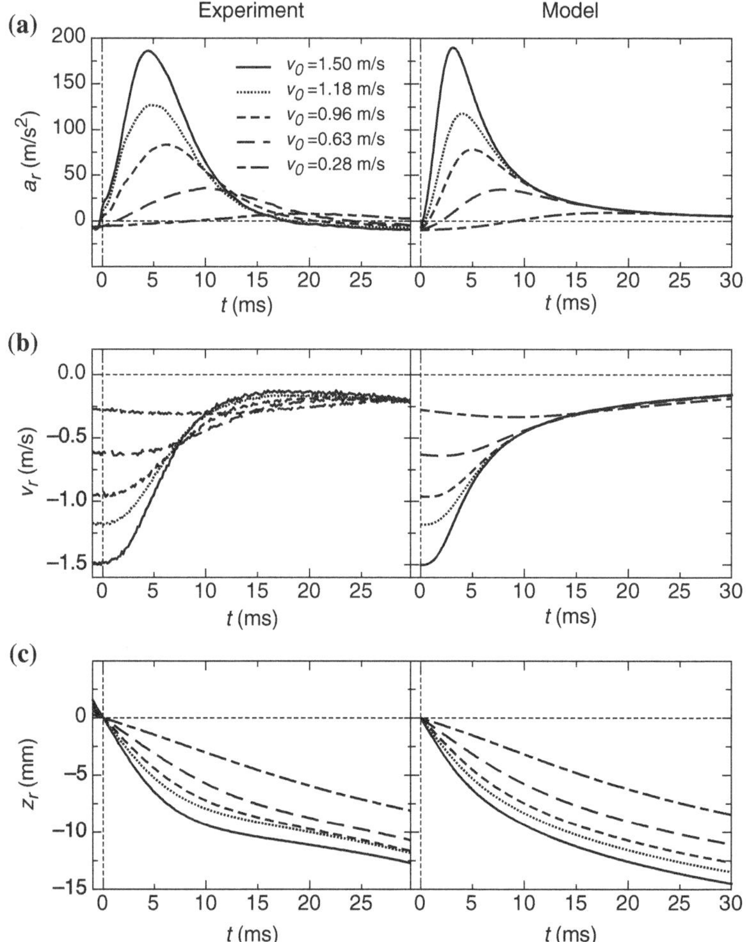

Fig. 2.15 Comparison of added mass model with experimental results. Experimental (*left column*) and numerical (*right column*) results for acceleration (**a**), velocity (**b**) and position (**c**) of rod impacting into suspension with $\eta = 1.0\,\text{cP}$, $\phi_0 = 0.49$, and impact velocities $v_0 = 1.50\,\text{m/s}$ (*solid line*), 1.18 m/s (*dots*), 0.96 m/s (*small dash*), 0.63 m/s (*large dash*) and 0.28 m/s (alternating *small/large dash*). Numerical results are Mathematica solutions to Eqs. 2.3 and 2.4 with parameters $m_r = 0.368\,\text{kg}$, $k = 12.5$, $r_r = 0.93\,\text{cm}$ and $\rho_s = 1295\,\text{kg/m}^3$ and initial conditions $v_r(0) = -v_0$ and $z_r(0) = 0$

show that the impact creates substantial flow in a region that extends $k|z_r|$ below and radially away from the rod. Approximating these points as bounding a cone-like region gives m_a the form:

$$m_a = C\frac{1}{3}\rho_s\pi(r_r + k|z_r|)^2 k|z_r|, \tag{2.4}$$

where ρ_s is the density of the suspension. Using this in Eq. 2.3 with the initial conditions $v_r(0) = -v_0$ and $z_r(0) = 0$ allows us to solve numerically for the rod dynamics. With the average measured value for the relative front growth rate ($k = 12.5$) and leaving the coefficient C as the only adjustable parameter, this minimal model reproduces the important features impact response surprisingly well over the whole range of initial velocities tested (Fig. 2.15). We find the best agreement for $C \approx 0.37$, similar to what is encountered for disk impact into regular liquids [4]. (In Appendix C, we extract m_a directly from our data and confirm the scaling with $|z_r|$ as given by Eq. 2.3. In Appendix D, we show how similar behaviors can arise if the growing solid column below the rod experiences viscous drag from the surrounding suspension.)

References

1. M. Shiffman, D.C. Spencer, The force of impact on a cone striking a water surface (vertical entry). Commun. Pur. Appl. Math. **4**(4), 379–417 (1951)
2. C.E Brennen, A review of added mass and fluid inertial forces. Report to the Deptartment of the Navy (CR 82.010) (1982)
3. A.A. Korobkin, V.V. Pukhnachov, Initial stage of water impact. Annu. Rev. Fluid Mech. **20**(1), 159–185 (1988)
4. J.W. Glasheen, T.A. McMahon, Vertical water entry of disks at low froude numbers. Phys. Fluids **8**, 2078–2083 (1996)
5. R. Bergmann, D. Van Der Meer, S. Gekle, A. Van Der Bos, D. Lohse, Controlled impact of a disk on a water surface: cavity dynamics. J. Fluid Mech. **633**, 381–409 (2009)
6. M. Ambroso, R. Kamien, D. Durian, Dynamics of shallow impact cratering. Phys. Rev. E **72**(4), 041305 (2005)
7. M. Hou, Z. Peng, R. Liu, K. Lu, C. Chan, Dynamics of a projectile penetrating in granular systems. Phys. Rev. E **72**(6), 062301 (2005)
8. H. Katsuragi, D.J. Durian, Unified force law for granular impact cratering. Nat. Phys. **3**(6), 420–423 (2007)
9. D.I. Goldman, P. Umbanhower, Scaling and dynamics of sphere and disk impact into granular media. Phys. Rev. E **77**(2), 021308 (2008)
10. S. Von Kann, S. Joubaud, G.A. Caballero-Robledo, D. Lohse, D. Van Der Meer, Effect of finite container size on granular jet formation. Phys. Rev. E **81**(4), 041306 (2010)
11. J.R. Royer, B. Conyers, E.I. Corwin, P.J. Eng, H.M. Jaeger, The role of interstitial gas in determining the impact response of granular beds. Europhys. Lett. **93**, 28008 (2011)
12. N.J. Wagner, J.F. Brady, Shear thickening in colloidal dispersions. Phys. Today **62**(10), 27–32 (2009)
13. D.R. Foss, J.F. Brady, Structure, diffusion and rheology of brownian suspensions by stokesian dynamics simulation. J. Fluid Mech. **407**(1), 167–200 (2000)
14. J.R. Melrose, R.C. Ball, Continuous shear thickening transitions in model concentrated colloids: The role of interparticle forces. J. Rheol. **48**, 937–961 (2004)
15. B.J. Maranzano, N.J. Wagner, Flow-small angle neutron scattering measurements of colloidal dispersion microstructure evolution through the shear thickening transition. J. Chem. Phys. **117**, 10291–10303 (2002)
16. D.P. Kalman, N.J. Wagner, Microstructure of shear-thickening concentrated suspensions determined by flow-USANS. Rheol. Acta **48**(8), 897–908 (2009)

17. X. Cheng, J.H. McCoy, J.N. Israelachvili, I. Cohen, Imaging the microscopic structure of shear thinning and thickening colloidal suspensions. Science **333**(6047), 1276–1279 (2011)
18. A. Fall, N. Huang, F. Bertrand, G. Ovarlez, D. Bonn, Shear thickening of cornstarch suspensions as a reentrant jamming transition. Phys. Rev. Lett. **100**, 018301 (2008)
19. E. Brown, H. Jaeger, Dynamic jamming point for shear thickening suspensions. Phys. Rev. Lett. **103**(8), 086001 (2009)
20. E. Brown, N.A. Forman, C.S. Orellana, H. Zhang, B.W. Maynor, D.E. Betts, J.M. Desimone, H.M. Jaeger, Generality of shear thickening in dense suspensions. Nat. Mater. **9**(3), 220–224 (2010)
21. E.E. Bischoff White, M. Chellamuthu, J.P. Rothstein, Extensional rheology of a shear-thickening cornstarch and water suspension. Rheol. Acta **49**(2), 119–129 (2010)
22. C.S. Orellana, J. He, H.M. Jaeger, Electrorheological response of dense strontium titanyl oxalate suspensions. Soft Matter **7**(18), 8023–8029 (2011)
23. A. Fall, F. Bertrand, G. Ovarlez, D. Bonn, Shear thickening of cornstarch suspensions. J. Rheol. **56**, 575–592 (2012)
24. E. Brown, H.M. Jaeger, The role of dilation and confining stresses in shear thickening of dense suspensions. J. Rheol. **56**(4), 875–924 (2012)
25. F. Merkt, R. Deegan, D. Goldman, E. Rericha, H. Swinney, Persistent holes in a fluid. Phys. Rev. Lett. **92**(18), 184501 (2004)
26. W.H. Boersma, J. Laven, H.N. Stein, Shear thickening (dilatancy) in concentrated dispersions. AIChE J. **36**(3), 321–332 (1990)
27. E. Brown, H. Zhang, N.A. Forman, B.W. Maynor, D.E. Betts, J.M. DeSimone, H.M. Jaeger, Shear thickening and jamming in densely packed suspensions of different particle shapes. Phys. Rev. E. **84**(3), 031408 (2011)
28. W.J. Frith, P. d'Haene, R. Buscall, J. Mewis, Shear thickening in model suspensions of sterically stabilized particles. J. Rheol. **40**, 531–549 (1996)
29. B. Liu, M. Shelley, J. Zhang, Focused force transmission through an aqueous suspension of granules. Phys. Rev. Lett. **105**(18), 188301 (2010)
30. S. von Kann, J.H. Snoeijer, D. Lohse, D. van der Meer, Nonmonotonic settling of a sphere in a cornstarch suspension. Phys. Rev. E **84**(6), 060401 (2011)
31. A.J. Liu, S.R. Nagel, Jamming is not just cool any more. Nature **396**(6706), 21–22 (1998)
32. L. Gómez, A. Turner, M. Van Hecke, V. Vitelli, Shocks near jamming. Phys. Rev. Lett. **108**(5), 058001 (2012)
33. L. Gómez, A. Turner, V. Vitelli, Uniform shock waves in disordered granular matter. Phys. Rev. E **86**(4), 041302 (2012)
34. M.D. Ediger, C.A. Angell, S.R. Nagel, Supercooled liquids and glasses. J. Phys. Chem. **100**(31), 13200–13212 (1996)
35. M. Papoular, Dense suspensions and supercooled liquids: dynamic similarities. Phys. Rev. E **60**, 2408–2410 (1999)
36. P.G. Debenedetti, F.H. Stillinger, Supercooled liquids and the glass transition. Nature **410**(6825), 259–267 (2001)
37. D.J. Graham, P. Magdolinos, M. Tosi, Investigation of solidification of benzophenone in the supercooled liquid state. J. Phys. Chem. **99**(13), 4757–4762 (1995)
38. M. Moghisi, P.T. Squire, An experimental investigation of the initial force of impact on a sphere striking a liquid surface. J. Fluid Mech. **108**, 133–146 (1981)

Chapter 3
Dynamic Jamming Fronts in a Model 2D System

3.1 Introduction

In the previous chapter, and in particular in Sects. 2.3 and 2.6, it was shown that the behavior of the suspension directly below the rod suggests that the rod motion causes it to "solidify" in the sense that this region moves rigidly with the rod and can bear stress and store elastic energy. We found that the vertical extent of this solidified region below the rod was proportional to the distance travelled by the rod below the original surface level, i.e. $h_f = kz_r^*$ (with the relative front growth rate $k \approx 12.5$ for $\phi_0 = 0.49$ in the cornstarch and water suspension). Loosely, we characterized this behavior as being reminiscent of a "snowplow." In this chapter, we focus on the physics of this snowplow solidification. We use a model 2D system of macroscopic disks that allows us to see the details of front formation on at individual particle level. We find that the snowplow solidification is intimately related to the concept of jamming. Although the jamming transition has been studied extensively in experiments and simulations [1–9], most studies have focused on the time-independent, bulk characteristics of the jammed state at fixed, uniform packing fraction. However, the front growth we are interested in here concerns dynamic features related to jamming when the packing fraction is not uniform in space or time. In the following chapter, we will use these results to predict how the front growth rate should depend on the cornstarch packing fraction in suspension.

3.2 Experimental Setup

The experiment consists of initially unjammed, binary-sized disks sitting on a plane that are forced toward jamming by uniaxial compression via a rake, as shown in Fig. 3.1. Laser-cut, black acrylic disks are randomly arranged at an initial packing fraction ϕ_0 on an acrylic tray that is backlit from below and recorded from above with a video camera. The disks are cut to sizes $d_l = 1.34$ cm and $d_s = 0.93$ cm

S.R. Waitukaitis, *Impact-Activated Solidification of Cornstarch and Water Suspensions*, Springer Theses, DOI 10.1007/978-3-319-09183-9_3

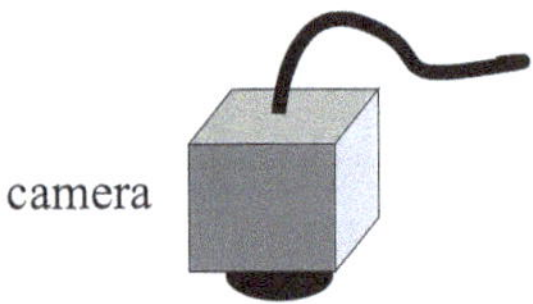

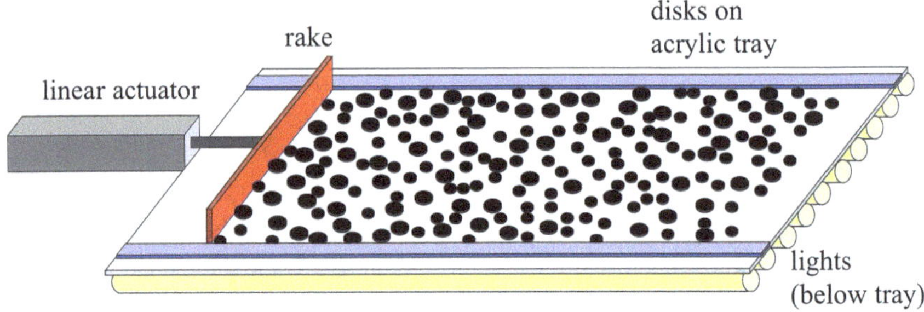

Fig. 3.1 2D model experiment for dynamic jamming. *Black* acrylic binary-sized disks (diameters $d_s = 0.93$ cm and $d_l = 1.34$ cm) in an initially unjammed state are compressed toward jamming by a rake. The rake is pushed at constant speed by a linear actuator. Flourescent lights shine through the opaque, *white* acrylic tray below the disks while the camera above records video

(i.e. $d_l/d_s \approx \sqrt{2}$) and are present in equal number $N_l = N_s$ in order to prevent crystallization [10–14]. The disks are then pushed from the left with a rake connected to a linear actuator that extends at constant velocity v_0. As can be seen from the images in Fig. 3.2, this leads to the formation of a stable densification front that travels ahead of the rake along x. Behind this front, the particles move with the velocity of the rake and are left in a jammed state with final packing fraction ϕ_J.

Frames from a video of the experiment are shown in Fig. 3.2. We analyze the videos by first binarizing and watershedding the images in ImageJ [15] in order to separate disks in close proximity. We determine the disk positions by locating all of the unique black domains (excluding the rake/actuator) within each image and calculating their centers of mass. Individual disk trajectories are acquired with the IDL tracking code developed by Crocker and Grier [16]. From these, we calculate the instantaneous disk velocities $v_{x,i}$ and $v_{y,i}$ by subtracting disk positions between frames and multiplying by the frame rate (usually ~10 frames per second). We use the Voronoi tessellation (obtained with Voro++ [17]) to calculate the local, instantaneous packing fraction of each disk ϕ_i (defined as the disk's area divided by the area of its Voronoi cell). In order to extract velocity and packing fraction profiles, we reduce the problem to one dimension and define the coarse-grained velocity field $V(x, t)$ and packing fraction $\phi(x, t)$ by binning the individual disk measurements at a given t along x (binsize $2d_l$) and calculating the bin averages. The variability of in ϕ_0 along these bins is ~0.01.

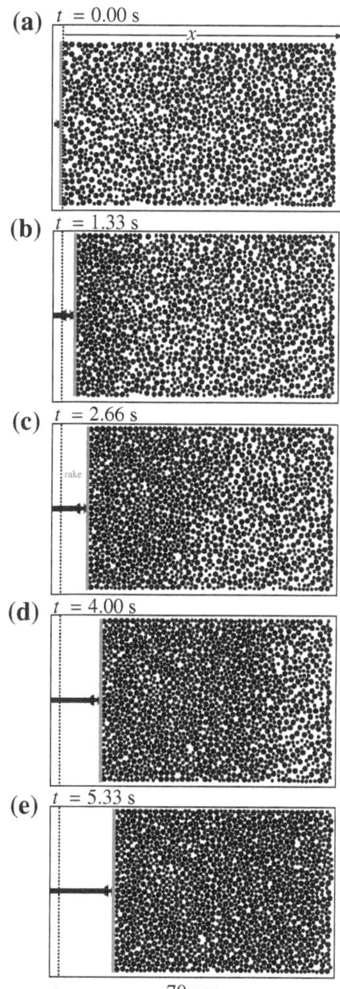

Fig. 3.2 Images from dynamic jamming. Images correspond to $\phi_0 = 0.611 \pm 0.003$, rake velocity $v_0 = 2.50\,\text{cm/s}$ at times $t = 0.00\,\text{s}$ (**a**), 1.33 s (**b**), 2.66 s (**c**), 4.00 s (**d**), and 5.33 s (**e**)

3.3 Characterization of the Relative Front Growth Rate

Figure 3.3 shows a snapshot of an initially uncompacted system (as in the right of the images) a short while after the rake has begun to move. By rendering images of the disks colored by their instantaneous velocity (Fig. 3.3a) and packing fraction (Fig. 3.3b), we are able to see the microscopic details of front formation. While disks near to the rake generally move with velocity v_0 and disks far ahead are stationary, we find velocities over the entire range $[0,\ v_0]$ in the transition region between these extremes. In the same region, one sees that the velocities are not constant along the transverse direction, but instead create fingerlike chains of particles that

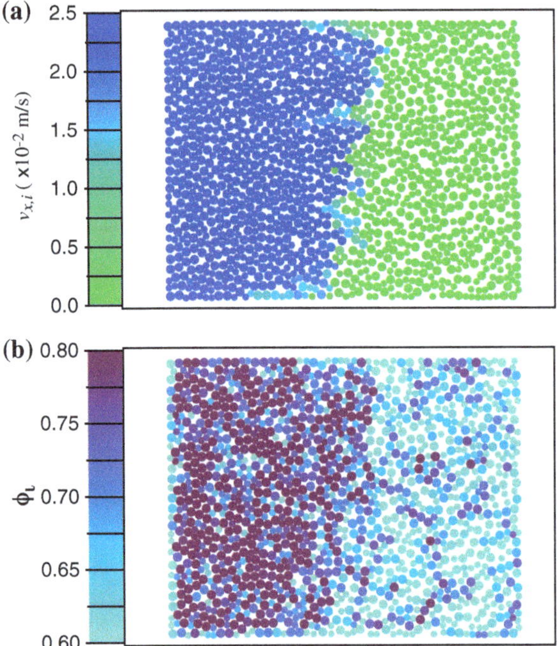

Fig. 3.3 Snapshot of a jamming front. **a** Rendering of disks colored by instantaneous axial velocity $v_{x,i}$ at $t = 4.00$ s in experiment with $\phi_0 = 0.611 \pm 0.003$ and rake velocity $v_0 = 2.50$ cm/s. **b** Same as (**a**) but with disks colored by local, instantaneous packing fraction ϕ_i

give roughness to the front. When we plot the coarse-grained velocity V versus x (Fig. 3.4a), however, these rough protrusions average into to a smooth profile with a soft transition from the rake velocity $V = v_0$ behind the front to $V = 0$ beyond the front. We find empirically that the these profiles are generally well-approximated by the equation

$$V(x) = \frac{v_0}{1 + e^{(x - x_f)/\Delta_f}}, \tag{3.1}$$

where x_f is the location of the center of the front and Δ_f is the width. We see similar profiles for $\phi(x)$ (Fig. 3.4b), which can be fit by the related equation $\phi(x) = \phi_0(1 - 1/(1 + e^{(x - x_f)})) + \phi_J$.

Fitting $V(x)$ to Eq. 3.1 allows us to extract measurements of the front position x_f and width Δ_f for each time t of the experiment. In Fig. 3.5a, we plot x_f versus t for several different values of ϕ_0 at a rake speed $v_0 = 1$ cm/s. As the plot shows, the fronts move at constant velocities, which we can measure by fitting to $x_f = v_f t$. Plotting v_f / v_0 versus ϕ_0 shows that the front speed appears to diverge as the packing fraction is increased. This kind of behavior arises from the requirement of disk conservation in combination with the fact that the disks cannot easily be compressed beyond

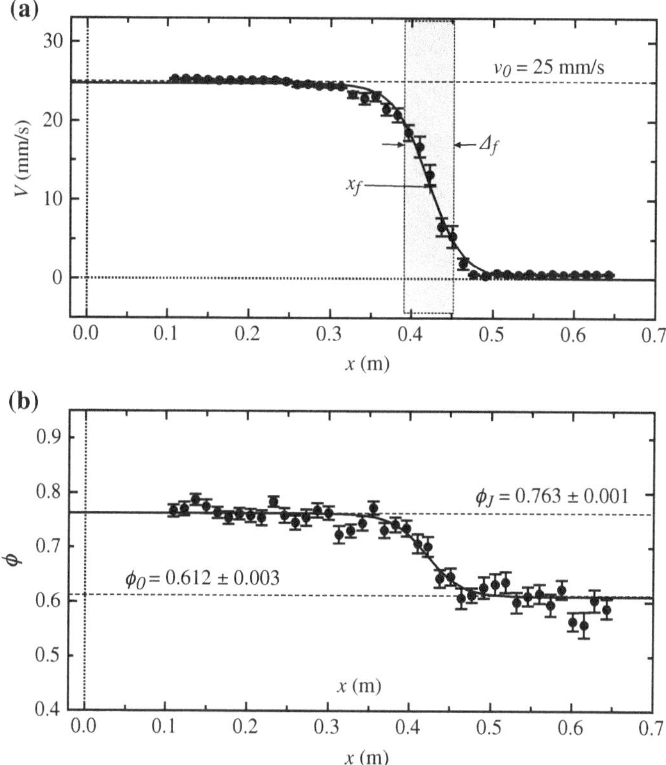

Fig. 3.4 Jamming front profiles. **a** Coarse-grained velocity V versus x at $t = 4.00$ s in experiment with $\phi_0 = 0.611 \pm 0.003$ and rake velocity $v_0 = 2.50$ cm/s (*black circles*). Curve is fit to Eq. 3.1, from which we extract the front position x_f and front width Δ_f (indicated in figure). **b** Profile of coarse-grained packing fraction ϕ versus x. Error bars in (**a**) and (**b**) are standard deviation of the mean for each x-bin

jamming. To see how this works, we start with the conservation equation

$$\frac{d\phi}{dt} + \frac{d}{dx}(V\phi) = 0. \tag{3.2}$$

As is often done [18], we assume an ad hoc constitutive equation for ϕ and V, in this case the simple linear relationship

$$\frac{\phi - \phi_0}{\phi_J - \phi_0} = \frac{V}{v_0} \equiv U, \tag{3.3}$$

where the dimensionless variable U scales between 0 and 1. This form effectively assumes that if the disks are at ϕ_J, they must also be moving at speed v_0, which is reasonable given that the step-like increase of the bulk modulus at jamming [5, 9, 19]

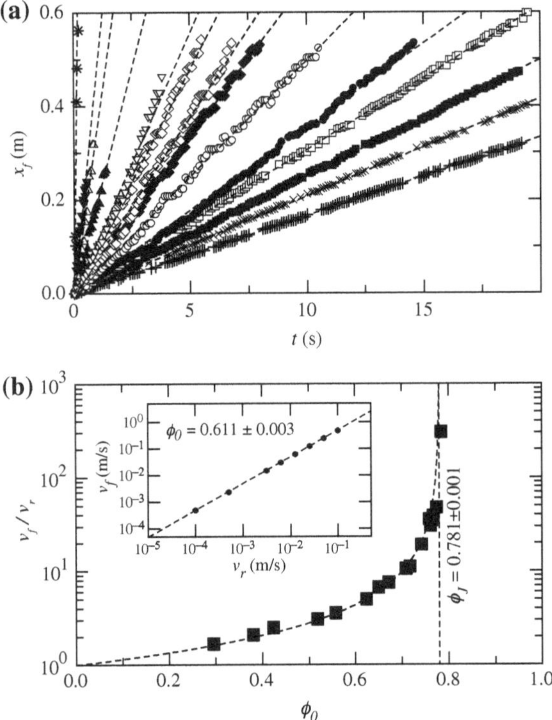

Fig. 3.5 Front trajectories. **a** Front trajectories x_f versus t for $v_0 = 1.0$ cm/s and with initial packing fractions (*bottom to top*) $\phi_0 = 0.296, 0.381, 0.423, 0.517, 0.557, 0.624, 0.650, 0.672, 0.694, 0.717, 0.741, 0.763, 0.774, 0.784$. **b** Ratio of front velocity to rake velocity v_f/v_0 versus ϕ_0. *Dashed line* is fit to Eq. 3.9 with fit parameter $\phi_J = 0.781 \pm 0.001$. Plot of v_f versus v_0 at fixed packing fractions shows $v_f \propto v_0$

will tend to prevent further compaction and cause disks to move rigidly with the rake. In Fig. 3.6, we confirm this is a very good approximation for our system. With Eq. 3.3, we can rewrite Eq. 3.2 in terms of U

$$\frac{d}{dt}\Big(U(\phi_J - \phi_0) + \phi_0\Big) + \frac{d}{dx}\Big(U v_0\Big(U + \frac{\phi_0}{\phi_J - \phi_0}\Big)\Big) = 0 \tag{3.4}$$

Which after a little rearrangement results in the Riemann formulation [20] of the inviscid Burgers equation,

$$\frac{dU}{dt} + \frac{d}{dx}\Big(U v_0\Big(U + \frac{\phi_0}{\phi_J - \phi_0}\Big)\Big) = 0. \tag{3.5}$$

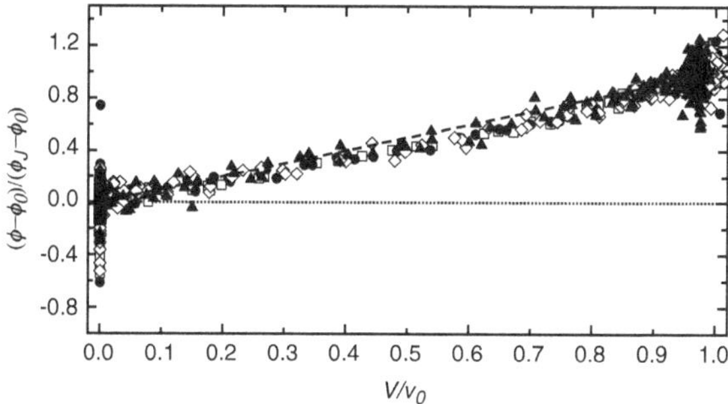

Fig. 3.6 Constitutive relation between packing fraction and velocity. Rescaled packing fraction $(\phi - \phi_0)/(\phi_J - \phi_0)$ versus rescaled velocity V/v_0 for dynamic jamming system with $v_0 = 1.0$ cm/s and initial packing fractions $\phi_0 = 0.428$ (*squares*), 0.619 (*circles*), 0.670 (*diamonds*), and 0.691 (triangles). Values for ϕ_J are calculated individually for each experiment

We can find the speed of the solidification front by using the boundary equations of our system in conjunction with the assumption that the solutions are traveling waves of the form $U(x,t) = U(x - v_f t) = U(X)$. This allows us to write both the spatial and time derivatives in terms of X, i.e.

$$v_f \frac{dU}{dX} = \frac{d}{dX}\left(U v_0\left(U + \frac{\phi_0}{\phi_J - \phi_0}\right)\right). \tag{3.6}$$

We now integrate this equation with respect to X,

$$v_f \int_{-\infty}^{\infty} \frac{dU}{dX} dX = \int_{-\infty}^{\infty} \frac{d}{dX}\left(U v_0\left(U + \frac{\phi_0}{\phi_J - \phi_0}\right)\right) dX \tag{3.7}$$

$$v_f(U|_\infty - U|_{-\infty}) = v_0\left(U\left(U + \frac{\phi_0}{\phi_J - \phi_0}\right)\bigg|_\infty - U\left(U + \frac{\phi_0}{\phi_J - \phi_0}\right)\bigg|_{-\infty}\right). \tag{3.8}$$

Finally, using the boundary conditions for our system $U|\infty = 0$ and $U|_{-\infty} = 1$, we arrive at the Rankine-Hugoniot jump condition [20] for the front speed,

$$v_f = v_0\left(1 + \frac{\phi_0}{\phi_J - \phi_0}\right). \tag{3.9}$$

As we pointed out in Chap. 2, the proportionality between v_f and v_0 is markedly different from the behavior of systems above jamming, where one encounters shock speeds that are either independent of the driving speed (weak shocks) or scale like $v_0^{1/5}$ (strong shocks) [21, 22]. Both above and below jamming, the distance to ϕ_J plays a critical role in determining the front speed. Here we see that the reason for this in systems below jamming essentially boils down to mass conservation, whereas with the shocks above jamming the dependence arose from the increase in the system pressure with packing fraction.

The relationship between the front velocity and rake velocity in Eq. 3.9 is closely related to the relative front growth rate discussed in Chap. 2. In particular, integrating Eq. 3.9 with respect to time yields

$$x_f = x_r\left(1 + \frac{\phi_0}{\phi_J - \phi_0}\right), \tag{3.10}$$

whre x_r is the displacement of the rake. This is identical to the relationship $h_f = |z_f| - |z_r| = k|z_r|$ encountered in Chap. 2, provided the relative front growth rate has the form

$$k = \frac{\phi_0}{\phi_J - \phi_0}. \tag{3.11}$$

This has the potential to explain both why the front growth in the suspension is proportional to the displacement of the rod and why the response is so sensitive to the particle packing fraction. In the following Chapter, we will test this predicted form for k with a speed-controlled impact experiment into a cornstarch and water suspensions.

For the system of disks here, fitting the data of Fig. 3.5b to Eq. 3.9 yields $\phi_J = 0.781 \pm 0.001$. It is clear from the fit, however, that there is a discrepancy at higher values of ϕ_0 as the fitted curve falls to the left of the data. As we are able to measure the final packing fraction for each experiment, we see that this actually arises because there is a trend such that higher ϕ_0 have higher ϕ_J (Fig. 3.7b). This variability may arise because of finite system size [5], but could also arise in part from the increased pressure at the front interface at higher v_0. Nonetheless, we achieve better agreement if we use the values of ϕ_J measured from each experiment individually. Doing so in conjunction with Eq. 3.9 accurately describes the front speed over nearly 2 decades in $\phi_J - \phi_0$ (note the dashed line in Fig. 3.7a is a prediction with no fit parameters). The range of values we find for ϕ_J (~0.76–0.79) is consistent with the results of recent simulations of frictional disks with identical size and number ratios [10, 11].

3.4 Anomalous Front Width

While disk conservation seems like the most natural starting point for modeling this system, it is incapable of predicting the fact that the fronts have finite widths (e.g. Fig. 3.3c). This is a robust feature of the system; the widths develop quickly after the rake begins to move and remain stable over time. Phenomenologically, this can be accounted for by the inclusion of diffusion, which results in the viscous Burgers equation given by

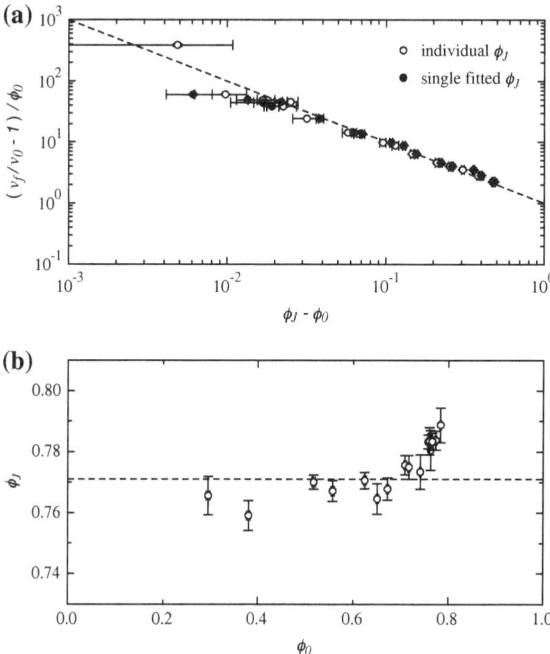

Fig. 3.7 Divergent front speed. **a** Front speed v_f versus rake speed v_0 on log–log scale for single, fitted ϕ_J (*solid circles*) and for individually measured ϕ_J (*open circles*). *Dashed line* is a prediction based on Eq. 3.9. **b** Measured final packing fraction ϕ_J versus initial packing fraction ϕ_0 shows slightly increasing trend

$$\frac{dU}{dt} + \frac{d}{dx}\left(U v_0 \left(U + \frac{\phi_0}{\phi_j - \phi_0}\right)\right) = D\frac{d^2U}{dx^2}, \tag{3.12}$$

where D is a diffusion coefficient (note that Eq. 3.1 is a solution). Equation 3.12 predicts that the width of the front is given by $\Delta_f = D/v_0$. In principle, the right hand side of Eq. 3.12 could be related either to viscous behavior, i.e. diffusion of momentum, or to the physical diffusion of particles. Dimensionally, D, which has units of length over time squared, can be interpreted as the product of a characteristic velocity and lengthscale. The only velocity scale in our system is the speed of the rake, which we have already shown affects the front speed, and, as we show in Fig. 3.8a, also sets the scale of the velocity fluctuations δV (see also Fig. 3.8b, which shows the fluctuations are insensitive to $\phi_J - \phi_0$). First guesses for the lengthscale might included the disk diameter, the tray width, or the mean free path. However, as we show in Fig. 3.9a, the width of the front dramatically *increases* as the system density is increased, while these lengthscales remain constant or approach zero.

On the other hand, recent simulations of driven systems just below the jamming transition have shown the presence of a divergent lengthscale associated with the

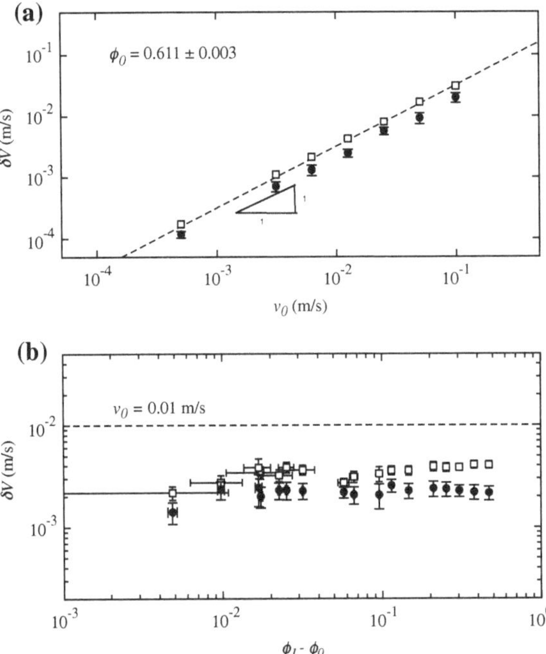

Fig. 3.8 Front velocity fluctuations. **a** Time-averaged longitudinal (*open squares*) and transverse (*closed circles*) root mean square velocity fluctuations δV at the front center versus rake velocity v_0. **b** δV versus $\phi_J - \phi_0$ show little dependence (symbols same as (**a**))

percolation of particle motion. In particular, O'Hern et al. showed that the lateral extent of moving particles via the insertion of an intruder diverges on approach to jamming [5]. (This result was confirmed in more detail by Reichhardt and coworkers [13]). More recently, Olssen and Teitel showed that the same-time transverse velocity correlation length for a sheared system of binary disks has similar divergent behavior [12]. In all cases, the authors found power law behavior for the lengthscale, i.e. $\xi \propto (\phi_J - \phi_0)^\nu$, with ν between ~0.6 and 0.7. (We mention that Vågberg et al. [23] found $\nu \approx 1$ in a similar binary-disk system by including finite size effects in the analysis. Here we use a more rudimentary analysis that should be compared with the exponents found by the other authors mentioned.) In Fig. 3.9b, we plot our Δ_f versus $\phi_J - \phi_0$ on a log-log graph to look for similar behavior. The data exhibit deviations from a straight line on the log-log graph, but if we assume they follow a power law, we see they are consistent with an exponent around 0.65, i.e. in the same range as the previous studies. This is suggestive that the growing front width we see here may be a signature of the divergent lengthscale seen more generally in jamming systems.

To test this directly, we measure the equal-time longitudinal velocity correlation function of our system. Symmetry dictates that the only position relative to which we

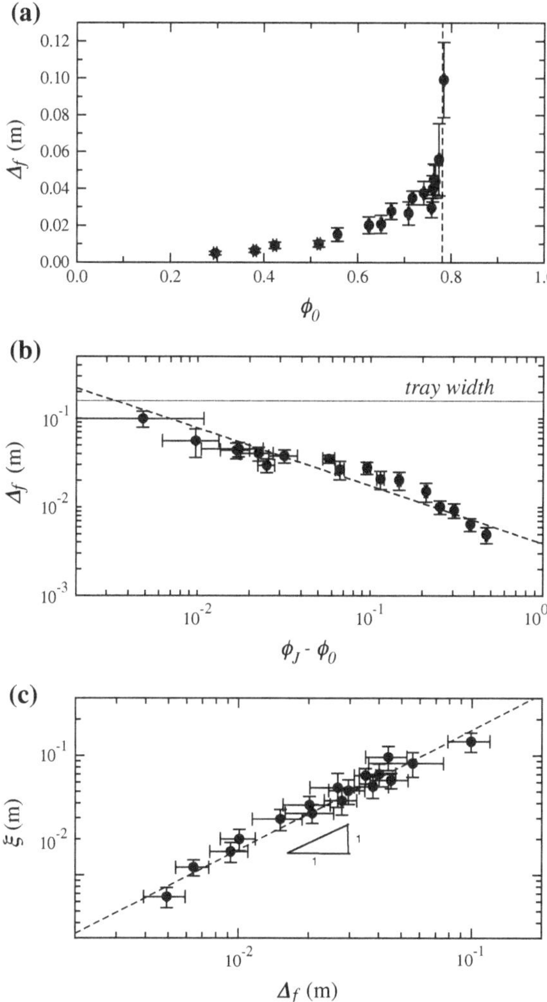

Fig. 3.9 Diverging front width. **a** Front width Δ_f versus ϕ_0 shows strong growth. **b** Δ_f versus $\phi_J - \phi_0$ on log–log graph (note individually measured values of ϕ_J are used). *Dashed line* is a power law with exponent -0.65 drawn as a guide to the eye (see text). **c** Longitudinal velocity correlation length ξ versus front width Δ_f. Fit is linear with coefficient ~1.6 and offset 0

can calculate a meaningful correlation function is at the center of the front (calculating it at positions behind the front will yield a correlation length that grows with time, and beyond it yields zero). We therefore define the correlation length here as

$$\xi = \frac{1}{v_0^2}\left(\int_{-\infty}^{x_f} V(x_f)V(x_f - x)dx + \int_{x_f}^{\infty} V(x_f)V(x - x_f)dx\right). \tag{3.13}$$

We do this calculation on both sides of the front because we find there is a slight tendency for the upstream value to be higher than the downstream value, indicating a slight asymmetry in the front shape. In Fig. 3.9c, we plot ξ versus Δ_f, which shows that they are indeed related by a simple linear scaling. If we look at Eq. 3.1 more closely, this shouldn't come as a surprise. Performing the above calculation (Eq. 3.13) with that profile shows that the correlation length and the width are related by the relation $\xi = \ln(2)\Delta_f$ (we actually measure a slope of ~1.6, very close to $2\ln(2) \approx 1.4$). This strongly suggests that the front width seen in dynamic jamming may be related to the diverging lengthscales seen in static jamming. Furthermore, while the experiments here concerned a 2D system of particles, the description is quite general and we expect similar, well-defined propagating fronts to develop generically in other systems, such as the cornstarch and water. We go on to discuss this more fully in this chapter.

References

1. A.J. Liu, S.R. Nagel, Jamming is not just cool any more. Nature **396**(6706), 21–22 (1998)
2. Z. Zhang, N. Xu, D.T.N. Chen, P. Yunker, A.M. Alsayed, K.B. Aptowicz, P. Habdas, A.J. Liu, S.R. Nagel, A.G. Yodh, Thermal vestige of the zero-temperature jamming transition. Nature **459**(7244), 230–233 (2009)
3. T.S. Majmudar, R.P. Behringer, Contact force measurements and stress-induced anisotropy in granular materials. Nature **435**(1079), 1079–1082 (2005)
4. D. Bi, J. Zhang, B. Chakraborty, R.P. Behringer, Jamming by shear. Nature **480**(7377), 355–358 (2011)
5. C.S. O'Hern, L.E. Silbert, S.R. Nagel, Jamming at zero temperature and zero applied stress: the epitome of disorder. Phys. Rev. E **68**(1), 011306 (2003)
6. A.S. Keys, A.R. Abate, S.C. Glotzer, D.J. Durian, Measurement of growing dynamical length scales and prediction of the jamming transition in a granular material. Nat. Phys. **3**(4), 260–264 (2007)
7. X. Cheng, Experimental study of the jamming transition at zero temperature. Phys. Rev. E **81**(3), 031301 (2010)
8. T. Majmudar, M. Sperl, S. Luding, R. Behringer, Jamming transition in granular systems. Phys. Rev. Lett. **98**(5), 058001 (2007)
9. A.J. Liu, S.R. Nagel, The jamming transition and the marginally jammed solid. Annu. Rev. Cond. Matter Phys. **1**(1), 347–369 (2010)
10. K. Shundyak, M. Van Hecke, W. Van Saarloos, Force mobilization and generalized isostaticity in jammed packings of frictional grains. Phys. Rev. E **75**(1), 010301 (2007)
11. S. Papanikolaou, C.S. O'Hern, M.D. Shattuck, Isostaticity at frictional jamming. Phys. Rev. Lett. **110**, 198002 (2012)
12. P. Olsson, S. Teitel, Critical scaling of shear viscosity at the jamming transition. Phys. Rev. Lett. **99**(17), 178001 (2007)
13. C.J. Olson Reichhardt, C. Reichhardt, Fluctuations, jamming, and yielding for a driven probe particle in disordered disk assemblies. Phys. Rev. E **82**(5), 051306 (2010)

14. J. Drocco, M. Hastings, C. Reichhardt, C. Reichhardt, Multiscaling at point j: jamming is a critical phenomenon. Phys. Rev. Lett. **95**(8), 088001 (2005)
15. C.A. Schneider, W.S. Rasband, K.W. Eliceiri, Nih image to imagej: 25 years of image analysis. Nat. Methods **9**(7), 671–675 (2012)
16. J.C. Crocker, D.G. Grier, Methods of digital video microscopy for colloidal studies. J. Colloid Interf. Sci. **179**, 298–310 (1996)
17. C.H. Rycroft, Voro++: a three-dimensional voronoi cell library in C++. Chaos **19**(4), 041111 (2009)
18. T. Beatus, T. Tlusty, R. Bar-Ziv, Burgers shock waves and sound in a 2d microfluidic droplets ensemble. Phys. Rev. Lett. **103**(11), 114502 (2009)
19. W.G. Ellenbroek, Z. Zeravcic, W. van Saarloos, M. van Hecke, Non-affine response: jammed packings versus spring networks. Europhys. Lett. **87**(3), 34004 (2009)
20. R. Courant, K.O. Friedrichs, *Supersonic Flow and Shock Waves* (Springer, Berlin, 1985)
21. L. Gómez, A. Turner, M. Van Hecke, V. Vitelli, Shocks near jamming. Phys. Rev. Lett. **108**(5), 058001 (2012)
22. L. Gómez, A. Turner, V. Vitelli, Uniform shock waves in disordered granular matter. Phys. Rev. E **86**(4), 041302 (2012)
23. D. Vågberg, D. Valdez-Balderas, M.A. Moore, P. Olsson, S. Teitel, Finite-size scaling at the jamming transition: corrections to scaling and the correlation-length critical exponent. Phys. Rev. E **83**(3), 030303 (2011)

Chapter 4
Speed-Controlled Impact into Cornstarch and Water Suspensions

4.1 Introduction

In Chap. 2, we performed experiments in which a rod was shot or allowed to fall via gravity into a cornstarch and water suspension and then freely accelerate in response to the forces acting on it. Our focus was on understanding why the rod slowed down, and we found evidence that this slowing was associated with the dynamic growth of a solid region below the impact site. In Chap. 3, we used a model system to show how the growth of a solid feature in a macroscopic system of hard disks is related to the jamming transition. In that system, we found that the relative front growth rate, k, strongly depends on the closeness to jamming $\phi_J - \phi_0$. If we assume that the suspension is made up of rigid cornstarch particles that are in an initially uncompacted state and that the interstitial fluid does not dramatically change the front behavior, it is easy to imagine that the same mechanism might account for the solid growth in suspension. (Although we do not know how to definitively validate these assumptions, we have taken steps to support them by showing that the cornstarch particles remain intact and hard after long term exposure to room temperature water (see Appendix E) and that, for a 1D system that exhibits dynamic jamming fronts, the only effect of a viscous interstitial liquid is to broaden the front (Appendix F).)

This proposition is in qualitative agreement with the fact that the peak acceleration during impact, a_{peak}, grows very rapidly with ϕ_0 (e.g. Fig. 2.4). However, freely accelerating impact is not the ideal situation for pulling out the details of front growth in the suspension. For one, the freely-impacting rod is continually slowing. Although the relative front growth rate of the disk experiment of Chap. 3 is independent of the push speed v_0, there is some speed dependence in the suspension (as we shall show shortly). In particular, the solid can "melt" before it reaches the bottom if the rod slows too much, or the rod can detach from the suspension surface if the impulse imparted to it is too great. In this chapter, we describe how we overcome these obstacles with speed-controlled impact into cornstarch and water suspensions. We use an Instron materials testing device, typically meant for stress-strain tests, to push a rod into the suspension surface with a constant velocity. Pushing the rod with a

S.R. Waitukaitis, *Impact-Activated Solidification of Cornstarch and Water Suspensions*, Springer Theses, DOI 10.1007/978-3-319-09183-9_4

constant velocity produces a much clearer signal to latch on to when the front hits the container bottom, leading to more precise measurements of k. We discover that in suspension, k is only independent of velocity for "fast push speeds," while at lower speeds the solidification vanishes. The crossover speed at which this occurs is set by the suspending liquid viscosity, with more viscous liquids enabling maximal front growth at smaller velocities. Ultimately, we find that the increase in the saturation values of k as a function of ϕ_0 is consistent with the results of Chap. 3 (i.e. Eq. 3.9), strongly suggesting that the impact thickening behavior in suspensions is intimately related to jamming.

4.2 Experimental Setup

We perform speed-controlled impact into cornstarch and water suspensions with an Instron 5586 materials testing device, as illustrated in Fig. 4.1. The Instron allows us to push a metal rod (radius $r_r = 0.93$ cm, as with the experiments of Chap. 2) into the suspension surface at a constant speed and simultaneously record its position z_r, velocity v_0, and the force exerted on it F_r. The suspension is again characterized by the suspending liquid viscosity η, the particle packing fraction ϕ_0, and the fill height H. We did not see viscosity or velocity dependence in the freely accelerating impact experiments, but the Instron allows (and constrains) us to operate at much slower speeds where the dependence on these parameters becomes apparent. (The maximum speed of the Instron is 8 mm/s, more than 25 times slower than the slowest speeds probed with the slingshot or free fall. We found it necessary to increase the suspension viscosity to work in this regime.)

Fig. 4.1 Speed-controlled impact experiment. Image showing how the Instron 5586 materials testing device pushes a metal rod into the surface of a cornstarch and water suspension at a constant velocity v_0 while simultaneously recording its instantaneous position z_r and the force exerted on it F_r. The suspension is characterized by its packing fraction ϕ_0, viscosity η, and fill height H. With the 500 N force transducer used with the Instron in these experiments, we had a force resolution of $\sim 2\times 10^{-3}$ N and a displacement resolution of 100 μm. Image of the actual experiment

Given that we performed many experiments at very slow push speeds (as low as 0.01 mm/s in some instances, leading to runtimes of over one hour for a single experiment), it was necessary to density match the suspending liquid with the cornstarch particles to prevent sedimentation. The process of mixing the suspending liquid was similar to that described in Chap. 2, but with additional steps for density matching. First, we mixed water with glycerin to achieve a liquid with the desired viscosity (exactly as in Chap. 2, Sect. 2). Next, we increased the density of this water/glycerin mixture by dissolving cesium chloride into it. The rough proportions necessary for this were extracted from the measurements taken by Johnson [1]. However, as data for the exact proportions of water, glycerin, and cesium chloride necessary to produce a desired viscosity and density were unavailable, we had to repeatedly add cesium chloride in small amounts and measure the density until we reached the desired result. For the data in this chapter, all suspending liquids had a density $\rho_l = 1590 \pm 10\,\text{kg/m}^3$.

4.3 Characterization of Force Curves

A typical result for the force-displacement curve resulting from a speed-controlled impact in which there is front formation is shown in Fig. 4.2. Before the rod hits the suspension surface no force is measured. Once it reaches the surface, the force

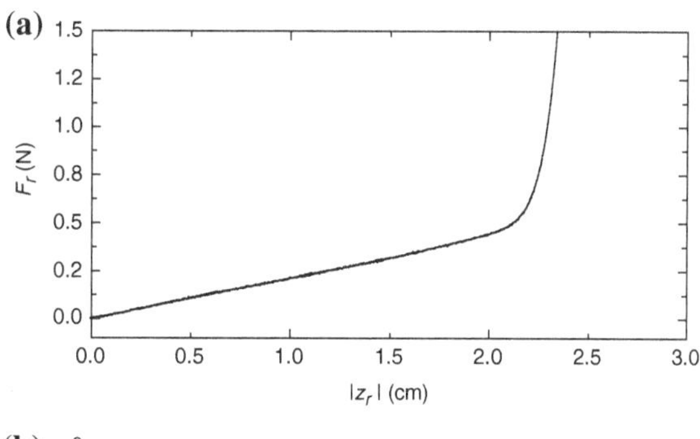

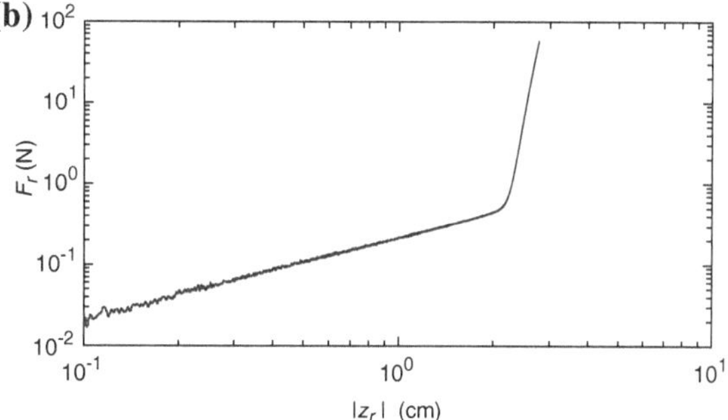

Fig. 4.2 Typical force-displacement curve. **a** Force on rod F_r versus distance below surface $|z_r|$ for suspension with $\eta = 14.0\,\text{cP}$, $\phi_0 = 0.50$, and $H = 4.8\,\text{cm}$ and $v_0 = 4.0\,\text{mm/s}$. The slowly rising initial region comes primarily from buoyant forces, while the quickly rising second region is caused by the jammed solid interacting with the container bottom. **b** Same as (**a**) but on log-log graph to show the strikingly different force scales in the two regimes

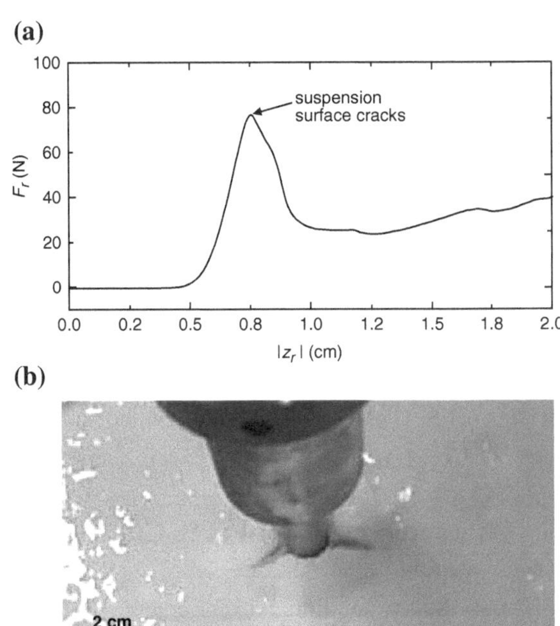

Fig. 4.3 Speed-controlled failure. **a** Full force-displacement curve for suspension with $\eta = 8.84\,\text{cP}$, $\phi_0 = 0.49$, $H = 4.46\,\text{cm}$ and $v_0 = 6\,\text{mm/s}$. The suspension fails after displacement $|z_r| \sim 0.75\,\text{cm}$, causing the force on the rod to plummet. **b** Image of the suspension surface shows that failure is associated with cracks that shoot radially outward from the rod

begins to increase slowly (e.g. before about 2.2 cm in Fig. 4.2a), growing at a rate of less than 1 N/cm. At this stage, although the jammed solid column is growing below the rod and the surrounding liquid is moving as well, the speed of the rod is so low that the added mass effect is small. [We can estimate from the added mass model of Chap. 2 (Eqs. 2.3 and 2.4) that the force caused by the growing solid at $|z_r| = 2.2\,\text{cm}$ in Fig. 4.2a is no more than $\sim$0.2 N, only about half of what we see in the figure.] Instead, the force here arises just as much from the buoyancy of the liquid displaced by the rod and surrounding depression. A little later, however, the force curve turns upward very quickly (after $|z_r| \sim 2.2\,\text{cm}$ in Fig. 4.2a), growing at a rate of $\sim$50 N/cm. As the zoomed out view in Fig. 4.2b shows, this sudden rise can lead to extraordinarily large forces and pressures on the rod (for the $r_r = 0.93\,\text{cm}$ rod, the maximum force is on the order of $\sim$100 N, equating to a maximum pressure of $\sim$0.5 MPa).

This is the signature of the solidified region of suspension interacting with the container bottom. If we allow the rod to keep pushing, the stress at the rod/suspension interface continues building and eventually the suspension "fails." An example of this is shown in the force curve of Fig. 4.3a, which shows a sudden drop off in the force after $|z_r| \sim 0.75\,\text{cm}$. Before this, the rod pushes the surface of the suspension downward rather than penetrate into it, as with the impact experiments (e.g. Fig. 2.2). After this, however, the rod penetrates significantly. The turnover in the force is accompanied by the sudden growth of cracks that shoot radially outward from the rod along the surface. These behaviors may be related to the cracking observed by

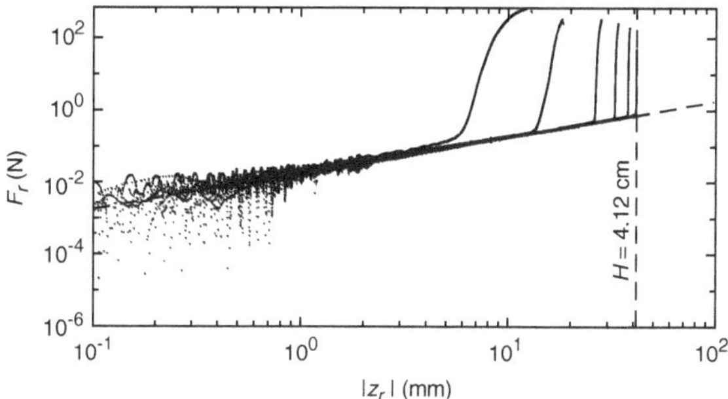

Fig. 4.4 Disappearance of front formation at low speeds. Force-displacement curves for speed-controlled impact into suspension ($\phi_0 = 0.48$, $\eta = 10.1$ cP, and fill height $H = 4.10$ cm) with push speeds (*left* to *right*) $v_0 = 8, 4, 2, 1, 0.4$, and 0.01 mm/s. The *dashed line* is the buoyant force arising from the weight of the suspension the rod displaces $\rho_s g \pi r_r^2 |z_r|$

Roché et al. [2] . While they are not our main focus here, this system provides a natural platform for studying the criterion for their onset.

Unlike freely accelerating impact, the control provided by the Instron allows us to push at speeds so slow that front formation ceases all together. In Fig. 4.4, we plot several force-displacement curves for increasingly slow speeds and show how this happens. At the highest push speed ($v_0 = 8$ mm/s), the effect of the bottom is seen when the rod has only travelled ~5 mm even though the fill height of the container is $H = 4.12$ cm. At this speed we see all of the visual features discussed previously, i.e. pushing rather than penetration and the growing depression around the impact site. As we slow the push speed, the story begins to change. In the force curves, the displacement at which the container bottom is felt becomes larger and larger until eventually at $v_0 = 0.01$ mm/s the front formation ceases all together and the uptick in F_r is felt only when the rod reaches the container bottom. On the suspension surface, these changes are accompanied by the rod beginning to penetrate into the suspension and the absence of the surface depression. In other words, the suspension begins to act more like a regular liquid. In doing so, the force on the rod (prior to the front hitting bottom) is also more liquid-like, arising solely from buoyancy. Following Archimedes principle, the dashed line in Fig. 4.4 is the weight of the suspension displaced by the rod $\rho_s g \pi r_r^2 |z_r|$, which shows this is the case.

4.4 Characterization of the Relative Front Growth Rate

Like the secondary peaks in the acceleration curves of the freely-impacting rod (see Figs. 2.8 and 2.9), the timing of the sudden increase in the force-displacement curves here changes with the suspension fill height H. This is shown in Fig. 4.5a, where we plot the curves for varying heights H with all other parameters held constant. As H is decreased, the upturn occurs at smaller $|z_r|$. In order to extract information

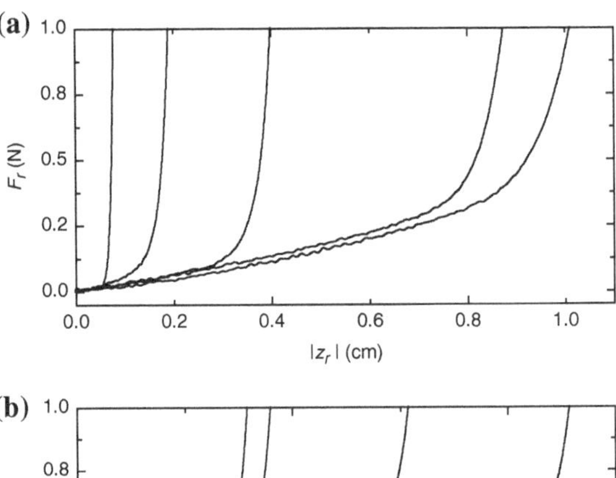

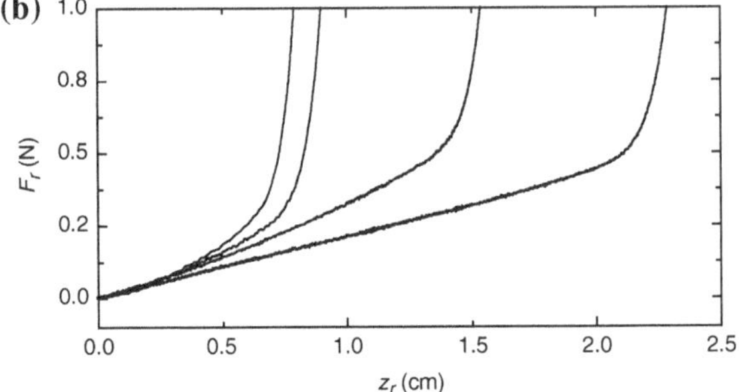

Fig. 4.5 Dependence of force-displacement curves on speed and fill height. **a** Force-displacement curves (*left* to *right*) for suspension with $\phi_0 = 0.49$ and $\eta = 12.4\,\text{cP}$, rod push speed $v_0 = 8.0\,\text{mm/s}$ and fill heights $H = 0.8, 2.5, 4.0, 6.5$, and $7.9\,\text{cm}$. **b** Force-displacement curves (*left* to *right*) for same suspension parameters as in (**a**), fixed height $H = 7.9\,\text{cm}$ and push speeds $v_0 = 1.0$, $2.0, 4.0$, and $8.0\,\text{mm/s}$

about the front growth from these data, we use a technique from solids testing that is commonly used to determine the yield stress of a material (although here what we are measuring is somewhat the opposite to yielding). The technique is illustrated in Fig. 4.6a, and consists of fitting both sides of the uptick in the force-displacement curve to lines and then determining where these fitted lines intersect. We define the position of the intersection as z_r^*. Similarly, we define the distance the front has grown during this time as $h_f = H - z_r^*$.

Being able to measure both the size of the jammed column h_f and the distance moved by the rod z_r^* allows to calculate the relative front growth rate $k = h_f/z_r^*$, as defined in Chap. 2 and discussed in the context of dynamic jamming in Chap. 3. For all k calculations in the rest of this chapter, we do this with a fixed suspension fill height $H = 8.5 \pm 0.1\,\text{cm}$. This value is chosen to be large enough to ensure that when the front forms, $h_f \gg r_r$, and to avoid using too much of the (prohibitively expensive) cesium chloride. In Fig. 4.7a, we show the speed dependence of k for a suspension with $\eta = 8.5\,\text{cP}$ and $\phi_0 = 0.49$. For high velocities (above $v_0 \sim 2\,\text{mm/s}$), k seems to come close to saturating. At smaller velocities, however, we see that k becomes smaller with decreasing velocity (consistent with the data in Fig. 4.4).

These measurements are qualitatively similar to the "melting" discussed at in the work of von Kann et al. [3]. While we do not at present fully understand the mechanism at play in this melting, we do know that it is related to the suspending liquid viscosity, as shown in Fig. 4.7b and c. For smaller viscosities, the k values at a given velocity are also smaller, presumably because the growing solid melts faster.

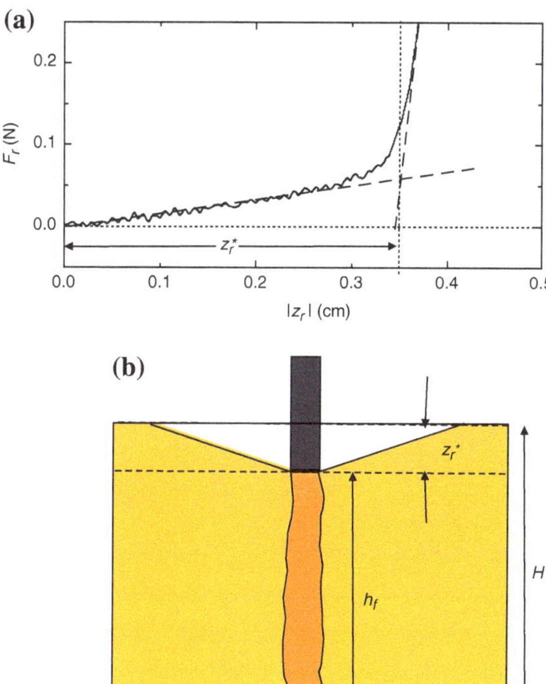

Fig. 4.6 Analysis of force-displacement curves. **a** Example force-displacement curve showing how the point of the upturn in force is determined by fitting the legs of the curve to two lines. The distance travelled by the rod at this point is defined as z_r^*. **b** Cartoon illustrating how the length of the jammed region is determined via the equation $h_f = H - z_r^*$

Comparing the curves for $\eta = 13.4\,\text{cP}$ (squares) and $\eta = 8.5\,\text{cP}$ (circles) shows that the saturation value for k is seemingly independent of the suspending liquid viscosity. (Presumably the data for $\eta < 8.5\,\text{cP}$ also saturate, but we were not able to go fast enough with the Instron to see this. However, the impact data from Chap. 2 with $\eta = 1\,\text{cP}$ water also suggest this is the case since we found the same value for k at $v_0 = 0.5$ and $1.0\,\text{m/s}$.) In Fig. 4.7c, we plot k versus the product of η and v_0, which collapses all the data onto a single curve. As the figure shows, the saturation value for this curve is consistent with what was seen in the impact experiments of Chap. 2, i.e. $k_{sat} \approx 12.5$. We can make a hand-waving argument for why the data collapse when plotted against ηv_0 by thinking of the "growth rate" of the solid as being proportional to v_0 while the "melting rate" is evidently proportional to $1/\eta$. The behavior transitions from primarily melting to primarily solidification when these two rates are equal, which leads to the product ηv_0.

Physically, k is a measure of the degree to which particles directly below the rod participate in solid growth. If k is small, no solid forms and the particles below the rod move laterally out of its way as would a normal liquid. If k is large, however, virtually all particles below the rod participate in front growth. This explains why k has a saturation value, which occurs when all the particles below the rod participate in solid growth. If this is correct, then the dependence of the saturation value of the relative front growth rate on the suspension packing fraction should be identical to the

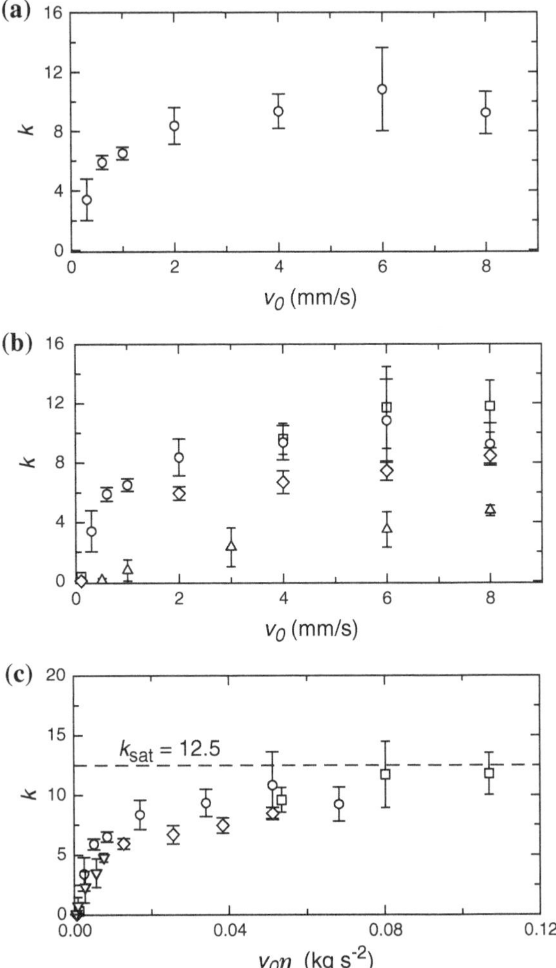

Fig. 4.7 Dependence of relative front growth rate on suspending liquid viscosity. **a** Relative front growth rate k versus impact velocity v_0 for a suspension with $\eta = 8.5$ cP and packing fraction $\phi_0 = 0.49$. **b** Same as (**a**) but with data for $\eta = 13.4$ cP (*squares*), 6.5 cP (*diamonds*), and 1.0 cP (*triangles*). **c** k data from (**b**) plotted against product of η and v_0 collapses to a single curve. *Dashed line* indicates the value of k found in Chap. 2 with the impact experiments

results encountered in Chap. 3 because the system, although it is 3D, actually reduces to a 1D problem in the region below the rod. In Chap. 3 (in particular Eq. 3.11), it was found that the relative front growth rate should scale with ϕ_0 like

$$k = \frac{\phi_0}{\phi_J - \phi_0}, \tag{4.1}$$

In Fig. 4.8 we test this idea by plotting all the k values we measured (without specificity to v_0 or η) versus ϕ_0. At a given packing fraction, changing the rod speed can vary k by more than an order of magnitude. Even so, the maximum values for different packing fractions, regardless of widely varying v_0 and η, do exhibit a clear

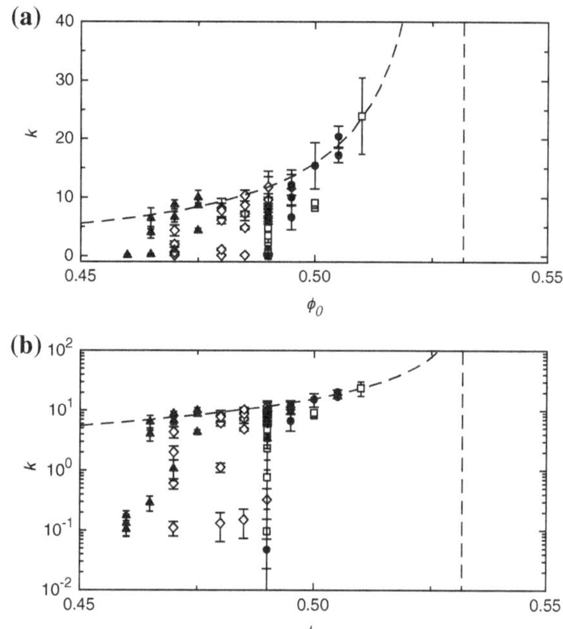

Fig. 4.8 Dependence of relative front growth rate on packing fraction. **a** Plot of relative front growth rate k versus packing fraction ϕ_0 for suspensions with suspending liquid viscosities $\eta = 80.1$ cP (I), 13.4 cP (*open diamonds*), 6.4 cP (*solid circles*) and 1.0 cP (*open squares*). The different points at each ϕ_0 value correspond to different speeds v_0. **b** Same as (**a**) but with log y-axis, showing that k varies by over 2 orders of magnitude. *Dashed line* in (**a**) and (**b**) is fit to saturation values at each ϕ_0 (excluding the data at the lowest ϕ_0) Eq. 4.1 with $\phi_J \approx 0.53$

trend. Although the range of ϕ_0 we were able to work with was limited, the saturation values of k in the region we did explore are consistent with Eq. 4.1 given that the packing fraction at which the suspension jams is $\phi_J \approx 0.53$. This number is comparable to but smaller than random loose packing for spheres and other values for random loose packing of cornstarch reported in the literature, which typically range from 0.55 to 0.57 [4, 5] (although those experiments are either carried out dry or with other suspending liquids). A simple explanation of this could be the uncertainty in the specific density of the grains. For example, if the density were 1.55 kg/l, as determined by Merkt et al. [6], ϕ_J could actually be as high as 0.54. Alternatively, this could also be caused be absorption of the suspending liquid by the starch (a worry of many authors [1, 6]), which would make the actual value of ϕ_J higher than what we report. It is also possible that an effect of the suspending liquid is to cause the suspension to seem to jam at lower values than it would otherwise.

Interestingly, we are not able to produce any front formation below $\phi_0 \sim 0.46$, no matter what viscosity of suspending liquid we use. This could arise for a variety of reasons. One possibility is that we simply did not use liquid that was sufficiently viscous to see the effect. (Note we could not make a proper suspension with viscosities beyond ~100 cP as the mixing became increasingly difficult and we could not rid the suspension of air bubbles, which actually seemed to cause it to shear thin.) Alternatively, if we believe the granular picture that our data support, this effect could arise from properties of the particle packing. For example, recent experimental work by Bi et al. with dry grains shows that there is a critical packing fraction *below* ϕ_J

associated with so-called shear jamming [7]. Concisely put, a shear jammed state is one in which a granular material with $\phi < \phi_J$ develops solid like properties, such as a bulk and shear modulus, as a result of the application of a finite amount of shear. This kind of behavior was seen by Seto et al. [8] in simulations of sheared suspensions that included frictional interactions between the particles, where the authors showed that discontinuous shear thickening only occurred beyond some critical packing fraction (also below ϕ_J). If the solid development arising from impact is associated in any way with shear jamming, this could explain why we see an onset packing fraction below ϕ_J at $\phi \sim 0.46$.

References

1. B. Johnson, The ultrasonic characterization of shear thickening suspensions, Ph.D. thesis, Washington University, May 2012
2. M. Roché, E. Myftiu, M.C. Johnston, P. Kim, H.A. Stone, Dynamic fracture of nonglassy suspensions. Phys. Rev. Lett. **110**(14), 148304 (2013)
3. S. von Kann, J.H. Snoeijer, D. Lohse, D. van der Meer, Nonmonotonic settling of a sphere in a cornstarch suspension. Phys. Rev. E **84**(6), 060401 (2011)
4. J.M. Valverde, A. Castellanos, Random loose packing of cohesive granular materials. Europhys. Lett. **75**(6), 985–991 (2006)
5. J.L. Willett, Packing characteristics of starch granules. Cereal Chem. **78**(1), 64–68 (2001)
6. F. Merkt, R. Deegan, D. Goldman, E. Rericha, H. Swinney. Persistent holes in a fluid. Phys. Rev. Lett. **92**(18), 184501 (2004)
7. D. Bi, J. Zhang, B. Chakraborty, R.P. Behringer, Jamming by shear. Nature **480**(7377), 355–358 (2011)
8. R. Seto, R. Mari, J.F. Morris, M.M. Denn, Discontinuous shear thickening of frictional hard-sphere suspensions (2013) arXiv:cond-mat.soft

Chapter 5
Results and Conclusions

The resistance of dense suspensions to intruding objects has typically been associated with shear thickening. However, there are a number of incompatibilities between existing shear thickening models and what occurs during impact. The most important inconsistency is that none of the shear thickening models can account for the stress scales observed in impact. In continuous shear thickening, the viscosity rise is just too mild. In discontinuous shear thickening, although the rise in viscosity can be very large, the stresses in the system are bounded by compliance of the weakest boundary, typically the liquid-air interface, limiting the stress to *at least* one order of magnitude smaller than what is required to keep a person running on the surface dry. Aside from this physical limitation, there are also a number of conceptual difficulties. For one, the prevalent models are only well-understood in the steady state, whereas impact is an inherently transient phenomenon. Additionally, these models were developed with strictly shear experiments, whereas impact involves both compressive and shear stresses.

A number of experiments studied suspensions under transient, compressive situations. While the split Hopkinson bar experiments led to stresses on the order of 50 MPa, these systems had no free boundaries. At least two experiments in the physics literature studied cornstarch and water suspensions in compressive situations with free boundaries, but these experiments did not study impact and instead worked with immersed spheres. Both found evidence for strong interaction between the immersed sphere and the container bottom. Although they interpreted this interaction as arising from a jammed region of suspension, their models failed to completely address the question of how this jammed region forms. By ignoring the jamming process, their models deemed the lower boundary as necessary for creating a large stress on the immersed sphere.

From the beginning, our approach has been to take a fresh look at this phenomenon and build an experiment that focused on studying impact directly. As a first step, we carried out a series of experiments in which we shoot a metal rod into a suspension of cornstarch and water. We see qualitative features of suspension impact that set it apart from impact into liquid or particles alone. With our high-speed camera focused

S.R. Waitukaitis, *Impact-Activated Solidification of Cornstarch and Water Suspensions*, Springer Theses, DOI 10.1007/978-3-319-09183-9_5

on the surface we see that, rather than penetrating into the suspension surface, the impacting rod pushes it downward as if it were a solid object. This occurs in conjunction with a rapidly growing surface depression that grows radially outward from the impact site. Furthermore, the absence of any splashing and the arrest of the suspension and rod motion once impact is over shows that this is a highly dissipative process.

These qualitative differences in what suspension impact looks like are accompanied by quantitative differences in the impactor dynamics. In particular, curves of the rod acceleration versus time in very deep suspensions exhibit smooth, well-defined peaks, which tell us that the force exerted by the suspension on the rod has both time-increasing and time-decreasing contributions. In agreement with our intuitive understanding of the material, we find that the scale of these acceleration peaks grows with impact velocity. For the highest speed impacts, this leads to pressures on the rod face exceeding 1 MPa, i.e. nearly 100 times larger than the upper bound in stress encountered in steady state discontinuous shear thickening experiments. Also in agreement with intuition, we find that these peaks become larger with increasing packing fraction. Surprisingly, however, they are insensitive to the suspending liquid viscosity and surface tension. The independence with respect to viscosity reemphasizes that suspension impact is not related to continuous shear thickening, where the suspension viscosity at a fixed packing fraction is simply proportional to the suspending liquid viscosity. The irrelevance of surface tension reemphasizes that the stress limitation in discontinuous shear thickening does not play a role during impact.

We probed for interactions between the impacting rod and the container bottom by changing the fill height H of the container. This leads to the emergence of secondary peaks in the rod acceleration versus time, indicative of the previously reported solid-like development in the suspension. We took this interpretation further by using our system to probe the growth of this solid region and in the process showing that a boundary interaction is not necessary for large compressive stresses. Our data show that the vertical extent of this solid region is proportional to how far the rod has pushed down the suspension surface. We defined the proportionality constant as the relative front growth rate k. With the aid of high-speed video of a laser sheet on the suspension surface to reveal the details of the depression growth and X-ray video to see the suspension interior's displacement field, we reconfirmed the proportional growth of this solid region and showed that is accompanied by a very large peripheral flow. Ultimately we were able to use the concept of added mass along with our measurements of the size of the moving region as a function of the rod displacement to make a simple model for how the rod is slowed during impact.

Suspecting the solid growth below the impacting rod might be related to jamming, we built a model 2D system comprised of macroscopic disks sitting on a plane and uniaxially compressed toward jamming with a rake. This simple system also produced solidification fronts whose extent beyond the impacting rake was proportional to the rake's displacement. We were able to show how this behavior is related to disk conservation in conjunction with the effective upper limit to the disk packing fraction at jamming. This allowed us to relate the relative front growth rate directly to the packing fraction. Interestingly, we also found that the widths of the jamming fronts in this system diverge on approach to jamming. This observation may be related to recent simulations of nearly identical disk systems that revealed a divergent lengthscale associated with the same-time velocity correlation function. The similarity suggests that some of the features characteristic of "static jamming" might have vestiges in dynamic jamming.

We were able to take what we learned from the model 2D system and look for similar packing fraction dependence in the 3D suspension. With speed-controlled impact experiments using essentially a stress strain test, we made more precise measurements of the relative front growth rate k as a function of both the packing fraction and suspending liquid viscosity. While the freely accelerating impact experiments of Chap. 2 suggested that the viscosity and velocity have no role in front development, the speed control of these experiments allowed us to see how these parameters come into play. We saw that at a fixed viscosity, k was constant at high v_0, but diminished to zero as v_0 was reduced. The location of this crossover changed with the suspending liquid viscosity, occurring at higher v_0 for less viscous suspensions. Additionally, we could collapse different k curves by plotting them against the product $v_0\eta$, suggesting that the "melting rate" of the suspension is proportional to $1/\eta$. Finally, we were able to show that the dependence of the saturation value of the relative front growth rate in suspension is consistent with the predicted form from the 2D experiments, provided that the cornstarch particles in suspension jam at $\phi_J \sim 0.53$.

In regard to the physics of suspensions, our work has a few major implications. First, we were able to show that it is the *transient* behaviors that lead to the interesting physics of dense suspensions during impact. While the typical approach in the past had been to drive suspensions until they reached a steady state response, it is now clear that we might gain a great deal of understanding from focusing on the behaviors before this stable response is reached. For example, it now seems very relevant to ask whether or not the viscosity of a sheared suspension depends on the shear strain (not just the shear rate, as has been previously assumed). Similarly, we can also wonder to what degree the properties of suspensions depend on system size, as we showed that the behavior of impact changed dramatically if the growing solid reached the lower boundary. Second, our work gives one of the first predictive connections between suspension behavior and jamming. Specifically, we showed that the growth of the saturation value of the relative front growth rate is consistent with the theoretical form we deduced from our dynamic jamming experiments with disks. Along with the proportionality between normal and shear stresses found in rheological measurements, this strongly encourages us to look at the physics of very dense suspensions with jamming in mind.

Outside of the physics of suspensions, our work points in a new direction for the jamming community, namely dynamic jamming. As we mentioned previously, most experiments in jamming have been concerned with static properties of the jammed state, such as the bulk modulus, shear modulus, or the specific value of ϕ_J. The jamming fronts we observe in suspension and in the 2D disk system show that there is a variety of fascinating behaviors associated with time-dependent process through which a jammed system is formed. We showed that the point at which a system jams can be determined experimentally by looking at the speed of jamming fronts. We also discovered an emergent lengthscale, the width of the jamming front, that diverges on approach to jamming. The growth of this lengthscale on approach to ϕ_J is reminiscent to other divergent lengthscales measured in simulations of static granular systems, strongly suggesting the two may be related. More work needs to be done to concretize this connection, but if shown to be the case this has the important implication that some of the elusive features predicted in the "static jamming" literature may have more easily accessible vestiges in dynamic jamming.

While we feel that the contributions we have made are important, we have uncovered just as many important questions along the way. In suspension, these mostly have to do with the role of the suspending liquid. Specifically, we wonder what sets the timescale for whether or not dynamic jamming occurs? Why do particles get caught up in the front rather than moving out of the way laterally, as a normal liquid would? What role does the liquid viscosity and particle packing fraction have on the jamming front width? What is the failure criterion that leads to penetration behavior in suspension? And why does the suspension seem to jam at a slightly lower value than might be expected? In regard to dynamic jamming itself, our questions are focused on the nature of the front width. Can we predict how the front width should grow with packing fraction? What changes would occur in the absence of interparticle friction? Or if the particle motion is not damped by inelasticity and external friction?

Answering these questions promises to be a lot of fun.

Appendix A
Penetration Regime in Freely Accelerating Impact

For sufficiently high speeds, the impact behaviors discussed in Chap. 2 change dramatically. In particular, rather than pushing the suspension surface downward and creating the surrounding depression, the rod penetrates into the suspension. As with the failure for speed-controlled impact discussed in Chap. 4, we also see cracks shoot out from the impact site, which may be related to the fracture seen by Roché et al. [1]. The penetration of the rod is especially evident in the surface depression measurements we make with the laser sheet, as in Fig. A.1, where we show how the $r_r = 0.93$ cm rod penetrates into the suspension surface when it hits with an impact speed of 4.0 m/s (note from Fig. 2.4b that the penetration transition occurs near $v_0 \sim 3.0$ m/s). Whereas the corresponding data for an impact velocity of $\sim$0.5 m/s in Chap. 2 showed a continuous color gradient across the rod-suspension boundary (indicating the rod did not penetrate), here we see a large difference in color across the boundary. At $\sim$8 ms, for example, the rod has cut into the suspension by nearly a full centimeter.

As discussed in regard to Fig. 2.4, we can see the transition to the penetration regime in the plots of a_{peak} and t_{peak} versus v_0. In Fig. A.2, we show how this behavior becomes more pronounced as the rod radius r_r is decreased. For the largest rod radius, $r_r = 0.93$ cm, we see that before $v_0 \sim 3.0$ m/s, t_{peak} is a decreasing function of v_0, while after this point it begins to curve upward slightly. At the same velocity, a_{peak} begins to grow less quickly with v_0. If we decrease the radius of the rod, this transition occurs at smaller v_0. Beyond the transition, the a_{peak} values for the smaller rods are smaller, while the t_{peak} values are larger. Interestingly, before the transition the curves for different radii are nearly the same.

Finally, we can also see the changes of the penetrating regime in the individual a_r versus t curves, as shown in Fig. A.3. For an impact in the pushing regime (e.g. Fig. 2.11, where $r_r = 0.93$ cm and $v_0 = 1.0$ m/s), we see the smooth characteristic peak discussed at length in Chap. 2. On the other hand, deep into the penetration regime (e.g. Fig. A.3, where $r_r = 0.21$ cm and $v_0 = 3.9$ m/s) , the peak becomes almost unrecognizable. Rather than a soft, parabolic shaped peak, we see slow growth to an extended peak followed by an abrupt fall off. This shows that the mechanism for the rod slowing in the penetration regime is very different, and likely much more complicated, than what we see in the pushing regime.

S.R. Waitukaitis, *Impact-Activated Solidification of Cornstarch and Water Suspensions*, Springer Theses, DOI 10.1007/978-3-319-09183-9

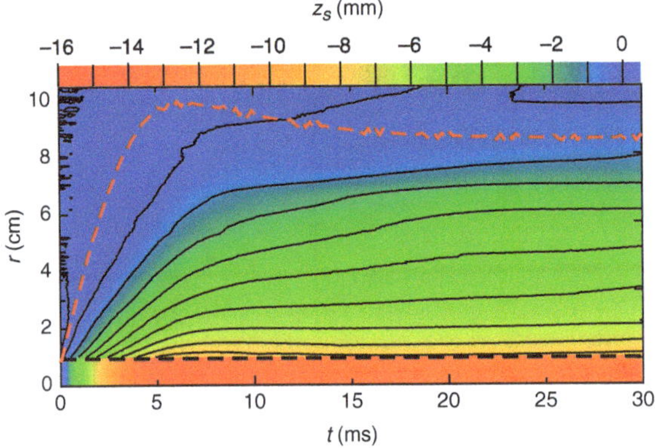

Fig. A.1 Space-time plot of surface depression in penetration regime. Depth of surface depression z_s (*color axis*) versus radial coordinate r and time t (with parameters $\eta = 1.0\,\text{cP}$, $\phi_0 = 0.49$, and $v_0 = 4.0 \pm 0.1\,\text{m/s}$). The *black dashed line* indicates the boundary between the rod and suspension. In contrast with Fig. 2.11, the discontinuous color scale across the boundary here shows that the rod penetrates significantly into the suspension

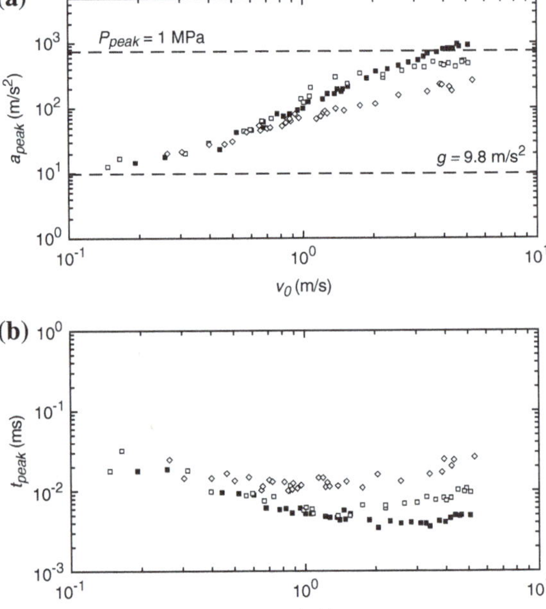

Fig. A.2 Increased penetration with smaller rod radii. **a** Peak acceleration a_{peak} versus impact velocity v_0 for rod radii $r_r = 0.93\,\text{cm}$ (*solid squares*), 0.48 cm (*open squares*), and 0.21 cm (*open diamonds*). Suspension has $\eta = 1.0\,\text{cP}$ and $\phi_0 = 0.49$. **b** Time to acceleration peak t_{peak} for same data as in (**a**)

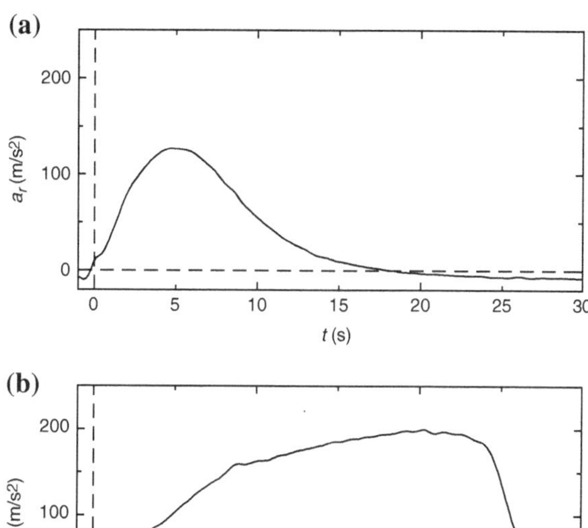

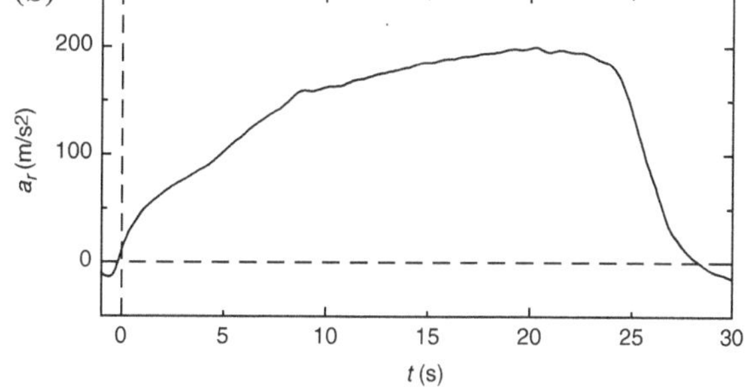

Fig. A.3 Changes to acceleration curves in penetration regime. **a** Rod acceleration a_r versus t for rod radii $r_r = 0.93\,\text{cm}$ and impact speed $v_0 = 1.0\,\text{m/s}$, i.e. in the pushing regime. Note the smooth, inverted parabola character to the peak. **b** a_r versus t for $r_r = 0.21\,\text{cm}$ and $v_0 = 3.9\,\text{m/s}$, i.e. deep in the penetration regime. The peak of the curve flattens out and then abruptly drops to zero

Appendix B
Details of X-ray Experiments

We captured X-ray video of the effect of impact on the suspension interior using a dental X-ray apparatus (Orthoscan High Definition Mini C-Arm, Model 1000-0004). The sensitivity of this system was too low to allow us to see changes in the packing fraction directly (as was possible with recent experiments involving the impact of a steel sphere into a granular bed using high-speed videography and high-luminosity synchrotron radiation [2]). Instead, we chose to investigate the suspension dynamics by loading it with metal tracer particles. These tracers were inserted by placing them in a line along the surface of the suspension below the rod and letting them sink into the field of view. We used a suspending liquid viscosity $\eta \sim 7\,\text{cP}$ to help the tracers sink slowly (recall changing the viscosity has little effect on the impact dynamics). The frame rate was limited to 30 frames per second, giving us "before" and "after" images of the impact with a typical time separation of $\Delta_t \sim 60\,\text{ms}$, as in Fig. B.1. We used particle imaging velocimetry (PIV) to calculate the local displacements between these images (Mathematica code written by Justin Burton). The algorithm we used had difficulty determining displacements near the suspension/air interface. However, the shape of the depression is easily seen from the images Fig. B.1b, and this allowed us to add "artificial" tracers (i.e. small squares of saturated pixels along the interface to help the algorithm in this region). The variability associated with letting the tracer particles sink into the field of view required us to take many videos in each field of view, reject PIV data from regions in which no tracer particles were present, and average the remaining results. The final data are the result of this procedure in four fields of view, the boundaries of which can be seen in Fig. B.1.

The X-ray apparatus was not powerful enough to image through the large ($30 \times 30 \times 30\,\text{cm}^3$) vat used for most of our experiments, and consequently we were forced to use a smaller ($10 \times 19 \times 30\,\text{cm}^3$) container. The nearness of the walls to the impactor in this container had a measurable effect on the motion of the rod, increasing the total impulse given to it. We show this in Fig. B.2, where we plot a_r, v_r, and z_r versus t for impact speeds $v_0 \approx 1\,\text{m/s}$ in the large and small containers. As the figure shows, the values of a_{peak} are similar for both containers, but the width of the peak is larger in the smaller container. This leads to a subtle bounce in the smaller container (note $v_r > 0\,\text{m/s}$ for $\sim 11\,\text{ms} \lesssim t \lesssim 28\,\text{ms}$) and a subsequent

S.R. Waitukaitis, *Impact-Activated Solidification of Cornstarch and Water Suspensions*, Springer Theses, DOI 10.1007/978-3-319-09183-9

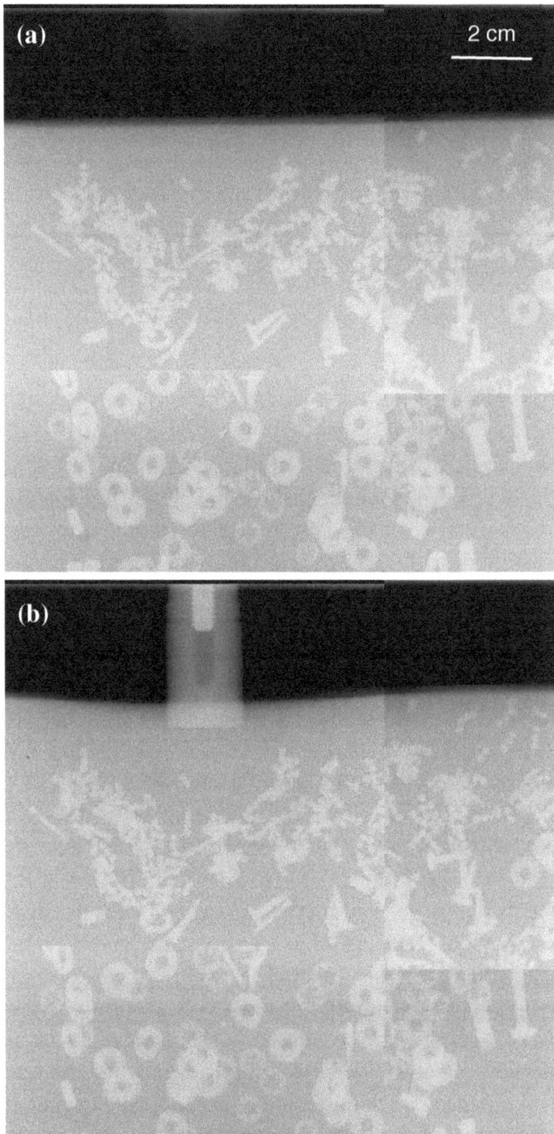

Fig. B.1 Before and after images from X-ray experiments. Tracer particles in suspension before (**a**) and $\sim$25 ms after (**b**) impact. Each image is composed from four fields of view. Although more suspension was imaged preferentially to the *right* of the rod, the container extends equally on both sides. PIV analysis between images such as these were averaged to obtain the impact displacement field

"pull" from the suspension as the rod moves upward ($a < -g$ starting at $t \sim 20$ ms). Despite these differences, the overall rod dynamics remain qualitatively similar, and the qualitative features of the displacement field from the smaller container are in agreement with what can be observed externally in the larger container (via the laser sheet measurements).

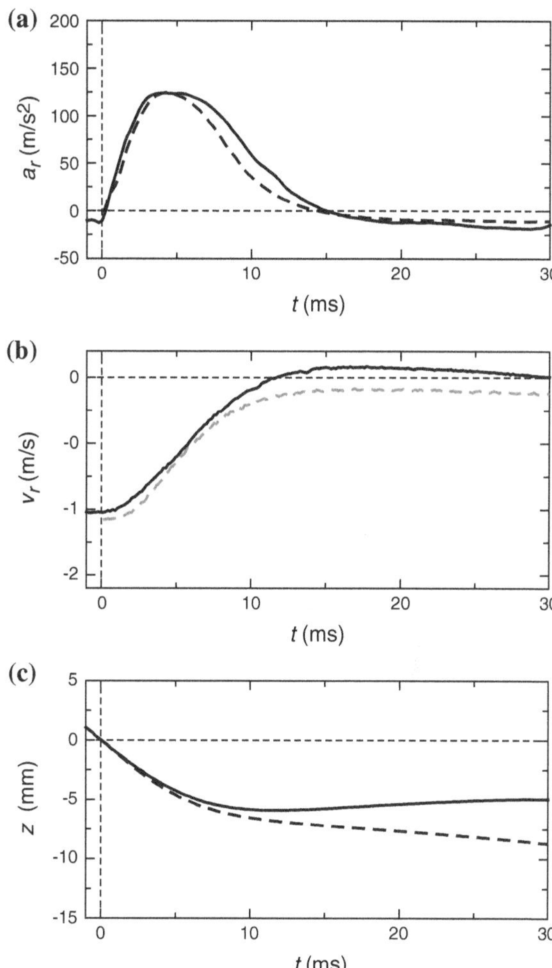

Fig. B.2 Changes in impact dynamics caused by smaller X-ray container. **a** Acceleration a_r versus time t for $30 \times 30 \times 30\,\text{cm}^3$ (*dashed line*) $10 \times 19 \times 30\,\text{cm}^3$ (*solid line*) containers. The small and large containers have similar values for a_{peak}, but the smaller container causes a_r to remain larger for an extended period of time. **b** v_r versus t with coloring same as in (**a**). The increased impulse leads to a subtle bounce of the rod ($v_r > 0\,\text{m/s}^2$ for $\sim 11\,\text{ms} \lesssim t \lesssim 28\,\text{ms}$). **c** z_r versus t with linestyles same as in (**a**)

Appendix C
Detailed Discussion of Added Mass

The collision of a solid object with a liquid is often modeled with the concept of added mass [3–5]. The key idea is to treat the loss of momentum of the impactor as resulting from an inelastic collision with an object whose mass m_a grows throughout the collision. This allows one to use Newton's laws and write the simple force equation

$$(m_r + m_a)\frac{d^2 z_r}{dt^2} = -\frac{dm_a}{dt}v_r + F_{ext}, \tag{C.1}$$

where F_{ext} accounts for all other forces not directly related to impact. Realistically, the situation is more complicated; the flowing region of liquid does not all move with one velocity but is instead varies smoothly with a velocity field. Nonetheless, the scale of the velocities involved in the liquid flow are set by the speed of the impacting object, and the added mass concept takes advantage of this by assuming constant liquid velocity (that of the impactor) over a finite region. The impact problem is therefore reduced to determining an appropriate form for the added mass. Typically, this will be a function of the geometry of the impactor and the distance it has moved since impact [4, 5].

In the case of suspension impact, the data from Chap. 2 (specifically Figs. 2.9 and 2.13) show that the suspension is primarily moving downward in a region extending $k|z_r|$ below the rod and $k|z_r|$ radially around the rod. Although this is caused by the formation of the solid column below the rod in the suspension, we can trick a regular liquid into behaving similarly. In Fig. C.1, we show how a wide-angle cone impacting into water or glycerine can lead to similar slowing via added mass. We choose an opening angle 2α such that the radial extent of the cone that is interacting with the liquid is the same as the extent of the depression in the cornstarch and water suspension (for $\phi_0 = 0.49$, as in Fig. 2.11a), i.e.

$$\tan(\alpha) = \frac{k|z_r|}{|z_r|} = 12.5. \tag{C.2}$$

Given the small changes in the liquid density and keeping the mass of the impactor fixed, the impact response of the cone into these liquids is strikingly similar to the

S.R. Waitukaitis, *Impact-Activated Solidification of Cornstarch and Water Suspensions*, Springer Theses, DOI 10.1007/978-3-319-09183-9

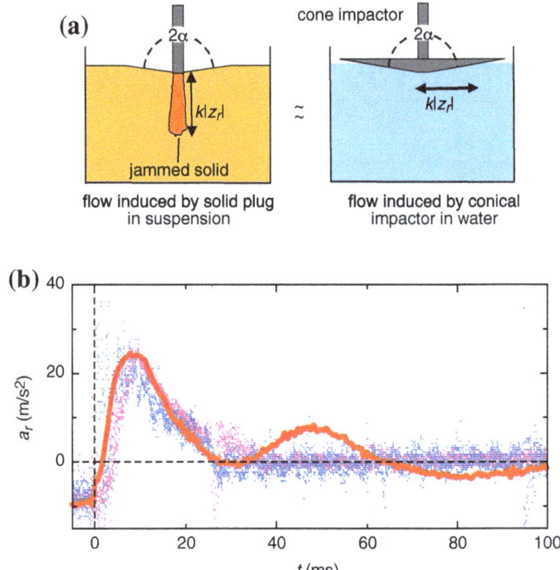

Fig. C.1 Added mass effect with cone impact into regular liquids. **a** Diagram illustrating similarity between solid plug induced flow field in the suspension and conical impactor induced flow field in a regular liquid. **b** Measurements of acceleration a_r versus time t for rod impact ($v_0 = 0.5$ m/s) into suspension (*red line*) compared to cone impact into water (*blue points*) and glycerol (*purple points*). The cone opening angle 2α is chosen to match that of the depression ($\tan(\alpha) = k = 12.5$)

rod into the suspension. Both produce peaks occurring at ∼7 ms which then slowly decay, and the heights of the peaks are both approximately 25 m/s^2. The only major difference is the existence of the second peak in the suspension impact, the signature of the solidified column reaching the container bottom. This of course is not apparent in the liquid experiments given they do not actually solidify.

If the force, displacement, and velocity of the impactor are known at all times, then we can extract the form of the added mass empirically from the impact data of Chap. 2. This is accomplished by using the momentum balance

$$(m_r + m_a)v_r = m_r v_0 + \int_0^t F_{ext} dt' \tag{C.3}$$

$$\rightarrow m_a = \frac{m_r v_0 + \int_0^t F_{ext} dt'}{v_r} - m_r. \tag{C.4}$$

We can take advantage of this to directly compare the added mass required to produce our acceleration curves to the form we have assumed in our model. In our experiment, the external forces are the gravitational pull on the rod $F_g = -m_r g$, the buoyant force of the liquid displaced in the depression $F_{buoy} \approx 1/3\pi\rho_s g(r_r + k|z_r|)^2|z_r|$, and any residual friction between the rod and guiderails F_f. Thus, we can extract the added mass from our experimental data via the equation

$$m_a = \frac{m_r v_0 - m_r g t + F_f t + \int_0^t \frac{1}{3}\pi\rho_s g(r_r + k|z_r|)^2|z_r| dt'}{v_r} - m_r. \tag{C.5}$$

We find that the frictional term is only important for the lowest velocity impacts ($v_0 < 0.5$ m/s), where releasing the rod from rest a few millimeters above the surface makes the rod susceptible to stick-slip just as it begins to move. We can estimate the scale of the friction by looking at the acceleration of the rod before $t = 0$, where it is generally constant and with a magnitude slightly smaller than g. In the $v_0 = 0.28$ m/s impact, for example, the friction is approximately constant with a magnitude of $1/3 m_r g$.

We plot m_a versus t for several v_0 in Fig. C.2a. The added mass grows more rapidly for higher v_0, but in all cases grows large enough to exceed the rod mass (horizontal dashed line) and far exceeds the maximum added mass that would be encountered in a regular liquid, $\sim 2/3\pi \rho_s r_r^3 \approx 2 \times 10^{-3}$ kg [6, 7]. If instead we plot

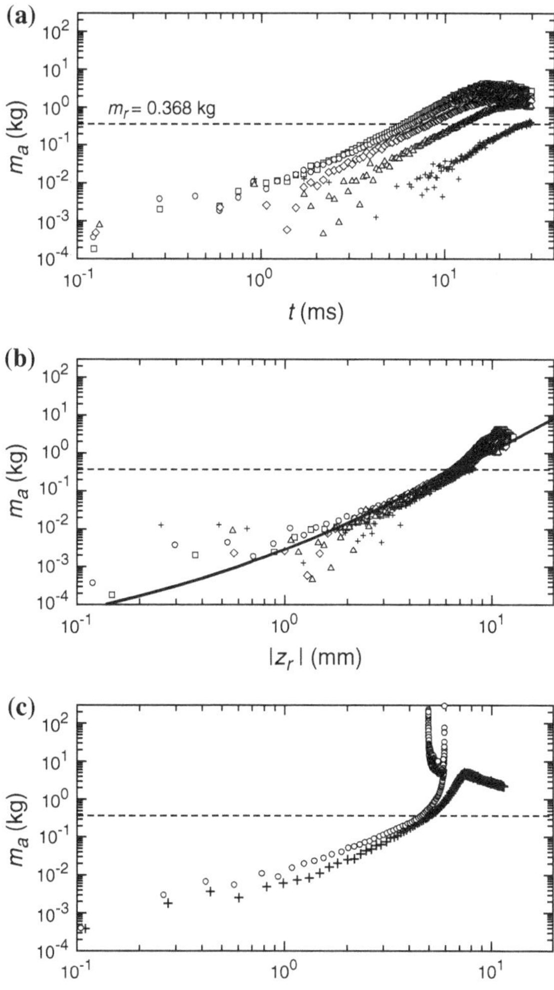

Fig. C.2 Emperically calculated added mass. **a** Added mass m_a versus time t for a $\phi_0 = 0.49$, $\eta = 1$ cP suspension with impact velocities $v_0 = 1.50$ m/s (*squares*), 1.18 m/s (*circles*), 0.96 m/s (*diamonds*), 0.63 m/s (*triangles*) and 0.28 m/s (*crosses*). The *horizontal dashed line* indicates the mass of the rod. **b** m_a versus $|z_r|$ with same color scheme as in (**a**). The *black line* is the form used in the model, i.e. Eq. 2.4. **c** m_a versus $|z_r|$ for impact speeds of ~ 0.5 m/s in the large ($30 \times 30 \times 30\,\text{cm}^3$, *circles*) and small ($10 \times 19 \times 30\,\text{cm}^3$, *crosses*) containers. The smaller container leads to a larger (divergent) spike in m_a at the end of the impact

m_a versus $|z_r|$, as in Fig. C.2b, the data collapse. Furthermore, the hypothesized form of the added mass (black line) closely follows the experimental data.

Toward the end of each impact, m_a undergoes a slight increase followed by an abrupt decrease. This is likely the signature of external forces that we have not accounted for or phenomena that cannot be encapsulated with added mass. One possibility is direct force transmission to the container boundaries, which would cause the added mass to increase abruptly as the impactor suddenly begins to push against infinitely large mass of the earth. Although we have tried to avoid this by making our container sufficiently deep, this can also arise if the flow approaches the sidewalls, which remains a possibility (note the radial extent of the depression is ~12.5 cm when $|z_r| = 1$ cm, leaving the wall a mere 2.5 cm away). Technical difficulties prevent us from using substantially larger containers to eliminate this effect, but we can verify it is present by showing it is much more pronounced in smaller containers, such as the one used for the X-ray measurements. In Fig. C.2c we plot the calculated added mass for impact speeds of $v_0 \approx 0.5$ m/s in the $30 \times 30 \times 30\,\text{cm}^3$ and $10 \times 19 \times 30\,\text{cm}^3$ containers, which shows that the increase in the added mass is much larger, and in fact diverges, in the smaller container (this can be explained by looking at Fig. B.2, where we see that v_r approaches zero and causes the empirically calculated m_a to diverge).

Appendix D
"Viscous" Model for Impact

The added mass model discussed in Chap. 2 and Appendix C shows how the slowing of the rod is associated with the transfer of momentum to a growing region of moving suspension. What is particularly insightful about this model is that it shows exactly why no boundaries are necessary for large impact response. At the same time, it is essentially a statement of conservation of momentum combined with empirical data relating the size of the flowing region to the motion of the rod. To gain a little more physical insight, we show in this appendix how similar dynamics (which are fully solvable) can arise if we assume that the force on the rod arises from viscous drag on the growing solid column below the impact site.

We begin by assuming that the jammed column below the impacting rod (whose extension below the surface is $h_f = k|z_r|$) experiences viscous drag from the surrounding suspension. On dimensional grounds, the drag force is

$$F_r = -\alpha\eta_s h_f v_r = \alpha\eta_s k z_r v_r, \tag{D.1}$$

where α is a coefficient of order unity. (Note we have assumed $z_r < 0$ and $v_r < 0$ as there can be a sign change if we assume otherwise.) In the early stages of impact when h_f is smaller than or comparable to than r_r, we expect that F_r might deviate from this simple form. However, as the front grows very quickly, $h_f < r_r$ only briefly. In principle the coefficient α may depend on r_r, but here we focus our attention on fixed r_r. This force is acting to slow both the rod and the solidified column of suspension. However, throughout the impact process the mass of the rod we use ($m_r = 0.368\,\text{kg}$) is much larger than the mass of the solidified column alone ($\pi r_r^2 h_f < 0.1\,\text{kg}$ if $h_f = 20\,\text{cm}$), therefore we can approximate the left hand side of Eq. D.1 as $F_r = m_r a_r$. Next, rewrite the equation as

$$m_r v_r \frac{dv_r}{dz} = \alpha\eta_s k z_r v_r. \tag{D.2}$$

S.R. Waitukaitis, *Impact-Activated Solidification of Cornstarch and Water Suspensions*, Springer Theses, DOI 10.1007/978-3-319-09183-9

We can integrate this equation to solve for the velocity as a function of the rod position, which gives

$$v_r - v_0 = \frac{\alpha k \eta_s}{2m_r} z_r^2, \tag{D.3}$$

This can be rewritten as the differential equation

$$\frac{dz_r}{dt} = v_0 + \frac{\alpha k \eta_s}{2m_r} z_r^2, \tag{D.4}$$

which we can integrate to find

$$\int_0^{z_r} \frac{1}{v_0 - \frac{\alpha \eta_s k}{2m_r} z_r'^2} dz_r' = \int_0^t dt'. \tag{D.5}$$

The solution to this equation is the inverse hyperbolic tangent. After some rearrangement, we find

$$z_r(t) = -\sqrt{\frac{2v_0 m_r}{\alpha k \eta_s}} \tanh\left(\sqrt{\frac{\alpha k \eta_s v_0}{2m_r}} t\right). \tag{D.6}$$

With this in hand, we differentiate to find the rod acceleration and velocity

$$v_r(t) = -v_0 \mathrm{sech}^2\left(\sqrt{\frac{\alpha k \eta_s v_0}{2m_r}} t\right) \tag{D.7}$$

$$a_r(t) = \sqrt{\frac{2\alpha k \eta_s}{m_r}} v_0^{3/2} \mathrm{sech}\left(\sqrt{\frac{\alpha k \eta_s v_0}{2m_r}} t\right)^2 \tanh\left(\sqrt{\frac{\alpha k \eta_s v_0}{2m_r}} t\right). \tag{D.8}$$

Working with the expression for $a_r(t)$, we can differentiate and set to zero to find expressions for peak a_{peak} and t_{peak}.

$$t_{peak} = \sqrt{\frac{m_r}{2\alpha k \eta_s v_0}} \mathrm{Log}(2 + \sqrt{3}) \tag{D.9}$$

$$a_{peak} = \sqrt{\frac{2\alpha k \eta_s}{m_r}} v_0^{3/2} \mathrm{sech}^2(\frac{1}{2}\mathrm{Log}(2 + \sqrt{3}))\tanh(\frac{1}{2}\mathrm{Log}(2 + \sqrt{3})). \tag{D.10}$$

To determine the appropriate coefficient α, we performed experiments in which we push the aluminum rod of radius $r_r = 0.93$ cm with constant velocity v_0 into the surface a high viscosity silicone oil (Clearco pure silicone fluid, $\eta = 9{,}77{,}000$ cP, $\rho_l = 0.977$ kg/L). Figure D.1a shows the force on the rod F_r as a function of the insertion depth $|z_r|$ for a few different velocities. As can be seen, the force on the rod grows linearly with $|z_r|$ for all velocities, but grows faster as v_0 is increased.

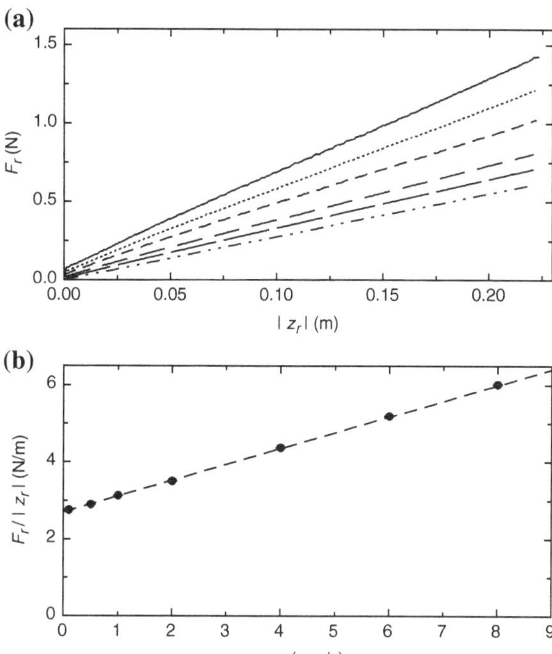

Fig. D.1 Viscous force on a submerged rod. **a** Force F_r versus distance of rod edge below surface $|z_r|$ for rod of radius $r_r = 0.93$ cm and silicone oil with density $\rho_l = 977\,\text{kg/m}^3$ and viscosity $\eta = 9{,}77{,}000$ Pa s. Curves have push speeds (*top* to *bottom*) $v_0 = 8.0, 6.0, 4.0, 2.0, 1.0, 0.5$ and 0.1 mm/s. **b** Slope of curves in (**a**) $F_r/|z_r|$ versus v_0. *Dashed line* fit is linear with offset 2.6 N/m and slope 410 Ns/m^2

Comparing the curves for $v_0 = 8.0$ and 4.0 mm/s, however, shows that the slope of these lines is not strictly linear with the velocity. This is because at these slow speeds we also have to account for the buoyant force on the submerged rod (an effect we can ignore during impact because $|v_r|$ is much higher). The full force on the rod in these measurements should be given by

$$F_r = \pi r_r^2 \rho_l g |z_r| + \alpha \eta |z_r| v_0, \tag{D.11}$$

which shows the slope of the curves, $F_r/|z_r|$, should have the form

$$F_r/|z_r| = \pi r_r^2 \rho_l g + \alpha \eta v_0. \tag{D.12}$$

In Fig. D.1b, we plot the slopes of the force displacement curves versus the push speed v_0. Fitting these data to Eq. D.12, we find an offset of 2.7 N/m (close to the predicted what we would predict from buoyancy $\pi r_r^2 \rho_l g \approx 2.6$ N/m) and a slope that is consistent with $\alpha = 0.42$.

With the prefactor α, we can now test the "viscous rod" model against the impact data in Chap. 2. In Fig. D.2, we again present the data for a_{peak} and t_{peak} versus v_0 for a suspension with $\phi_0 = 0.49$ and $\eta = 1.0$ cP (as in Fig. 2.4). The dashed lines are fits to Eq. D.10, which show that the growth a_{peak} is consistent with the scaling $v_0^{3/2}$ as is t_{peak} with the scaling $v_0^{-1/2}$ (in the velocity range 0.3–3.0 m/s, i.e. after

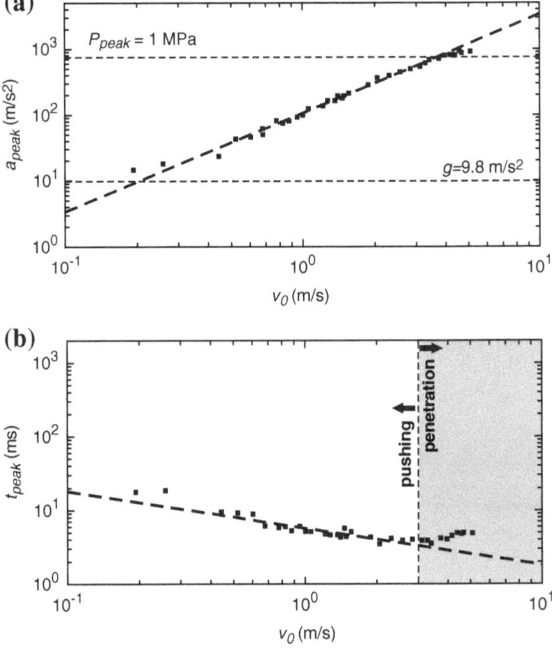

Fig. D.2 Peak scalings in "viscous" model. **a** Peak acceleration a_{peak} versus v_0 for suspension with $\phi_0 = 0.49$ and $\eta = 1.0$ cP. Fit is $\propto v_0^{3/2}$, as in Eq. D.10, with free parameter $\eta_s = 2650 \pm 60$ Pa s. **b** Time to peak t_{peak} versus v_0 for same suspension parameters as in (**a**). Fit is $\propto v_0^{-1/2}$, as in Eq. D.10, with fit parameter $\eta_s = 1{,}900 \pm 100$ Pa s. Fits in **a** and **b** are only performed with data for 0.3 m/s $< v_0 <$ 3.0 m/s

gravity becomes irrelevant but before the rod begins to penetrate). Given we know ρ_s, k, and α, the only free parameter in these fits is the effective viscosity of the suspension η_s. Fitting for both t_{peak} and a_{peak} separately yields an average value $\eta_s = 2260 \pm 390$ Pa s (note we get values $\eta_s = 2650 \pm 60$ Pa s and 1900 ± 100 Pa s for the individual a_{peak} and t_{peak} fits, respectively).

As we did with the added mass model in Chap. 2, we present the a_r, v_r, and z_r versus t curves for the experiment next to the results for the "viscous" model in Fig. D.3. We go beyond the simple law in Eq. D.1 (which we kept simple for the sake of determining scaling relations) to also include the effect of gravity on the rod, the buoyant force from the depression, and the added mass effect arising from the growing column, giving rise to a full force law

$$(m_r + \pi \rho_s r_r^2 k|z_r|)\frac{d^2 z_r}{dt^2} = \alpha k \eta_s z_r \frac{dz_r}{dt} - m_r g + \frac{1}{3}\pi \rho_s g |z_r|(r_r + k|z_r|)^2. \quad \text{(D.13)}$$

We use the previously calculated values for all parameters ($\eta_s = 2260$ Pa s, $\alpha = 0.42$, and $k = 12.5$) and solve the equation numerically. As can be seen, this compares perhaps even more favorably than the added mass model, reproducing very well both the height and the timing of the peaks. The model only seems to fail at large t, where the slowing is overestimated. This may be caused either by the melting of the solid, which would tend to reduce all slowing forces, or by disconnection of the rod from the suspension surface (which would have a similar effect). We remark

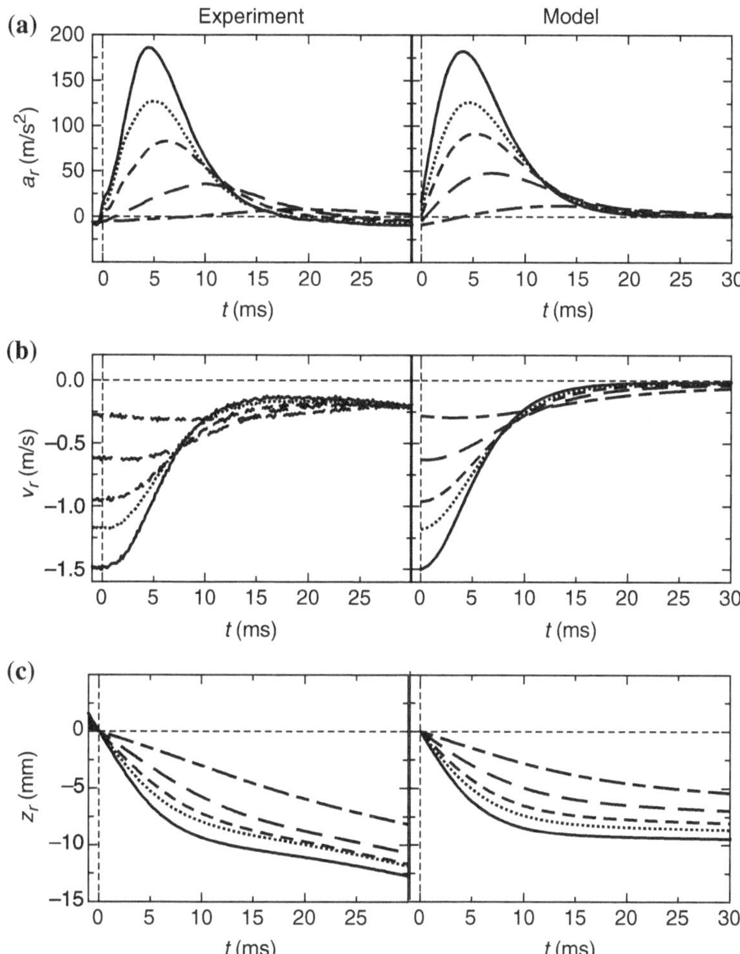

Fig. D.3 Comparison of "viscous" model with experimental results. Experimental (*left column*) and numerical (*right column*) results for acceleration (**a**), velocity (**b**) and position (**c**) of rod impacting into suspension with $\eta = 1.0\,\text{cP}$, $\phi_0 = 0.49$, and impact velocities $v_0 = 1.50\,\text{m/s}$ (*solid lines*), 1.18 m/s (*dots*), 0.96 m/s (*small dash*), 0.63 m/s (*large dash*) and 0.28 m/s (alternating *small/large dash*). Numerical results are Mathematica solutions to Eq. D.13 with parameters $m_r = 0.368\,\text{kg}$, $k = 12.5$, $r_r = 0.93\,\text{cm}$, $\rho_s = 1295\,\text{kg/m}^3$, $\alpha = 0.42$, $\eta_s = 2260\,\text{Pa s}$ and initial conditions $v_r(0) = -v_0$ and $z_r(0) = 0$

that the value for η_s is surprisingly close to the "geometric viscosity" of similarly packed cornstarch and water suspensions as measured just before the system reaches τ_{max} [8]. Interestingly, the data in Fig. 2.4 suggest this effective viscosity is still independent of the suspending liquid viscosity. This could be a consequence of the particle interactions transitioning from liquid (viscous) dominated to particle (friction) dominated, as proposed by [9] and seen in simulations by Seto et al. [10].

Appendix E
Cornstarch Particle Modulus

Much of our work suggests that when cornstarch particles are immersed in water, they remain essentially hard particles that interact primarily through steric repulsion. To test this proposition, and noting that there is little literature on the physical properties of cornstarch as a material (aside from the food science literature, which tends to focus on its gelatinization at higher temperatures [11–13]), we decided to study them using an atomic force microscope (AFM, Veeco Multimode 8 with ScanAsyst). This not only allowed us to take a detailed look at the surface morphology of the particles, it also enabled us to measure the material's Young's modulus. We performed measurements on dry particles, but we also looked for softening effects by soaking the particles in water overnight ($\sim$12 h), blowing the excess water off the particles, and then quickly putting them under the microscope.

Figure E.1 shows a schematic of the measurement operation. We glue individual cornstarch particles to a microscope slide. Then, we probe the particle surface by bringing the AFM cantilever (Tap525A, P/N MPP-13120-10, nominal spring constant $\sim$200 N/m) into contact with the particle surface, which causes it to deflect. From the deflection of the laser on the upper side of the cantilever, we are able to measure both the vertical position of the cantilever as well as the force exerted on it as it pushes into the cornstarch particle.

Figure E.2 shows a typical plot of the force on the AFM cantilever F_c as a function of the displacement of the cantilever from the particle surface δ_c ($\delta_c > 0$ indicates the cantilever is above the surface). During the approach (black curve), virtually no force is felt until the cantilever is within 2 nm of the surface, at which point it feels a slight attraction. Once it begins to physically press against the surface, it feels a strong repulsive force, which is also seen on the withdraw from the surface. Finally, as it is pulled away, it feels a slight "stickiness" as the surface tries to hold onto it.

In practice, these force-displacement curves are taken with respect to an arbitrary vertical lab coordinate. The location of the surface and transition the variable δ_c is determined from the curve as the average position of the minima of the approach and retraction curves. Determining the height of the surface this way while scanning the cantilever over a large area of the particle allows us to generate a height profile of the surface, a typical example of which is shown in Fig. E.3. These data show that the

S.R. Waitukaitis, *Impact-Activated Solidification of Cornstarch and Water Suspensions*, Springer Theses, DOI 10.1007/978-3-319-09183-9

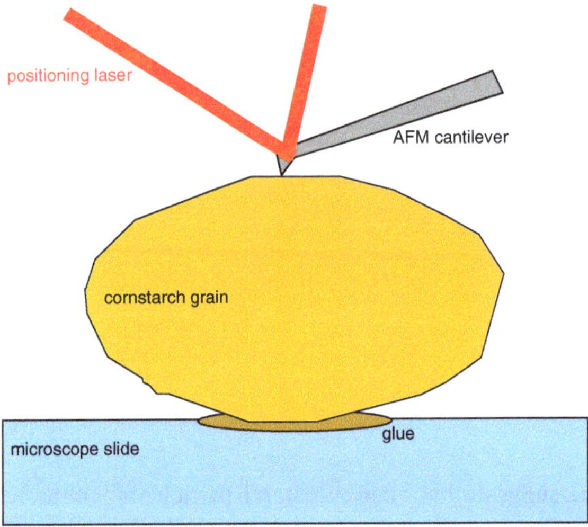

Fig. E.1 Probing cornstarch as a material with an AFM. Individual cornstarch particles are glued to a microscope slide. We measure their surface properties by poking them with an AFM cantilever, which gives us high spatial frequency access to the particle height profile and local modulus. Note this figure is not to scale

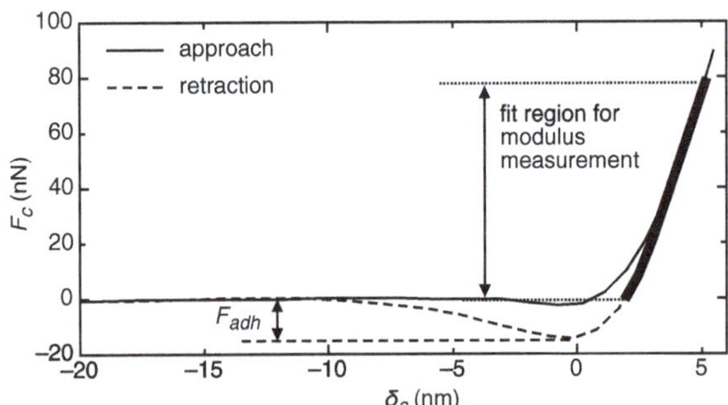

Fig. E.2 Typical AFM force-displacement curve. Force measured on AFM cantilever F_c versus distance of cantilever from particle surface δ_c. The *solid curve* corresponds to the approach toward the particle, while the *dashed curve* corresponds to the retraction. The thick *black line* on the retraction curve indicates where the fit to extract the Young's modulus is performed (Eq. E.1). The adhesive "sticking force", F_{adh}, is also indicated

cornstarch particles are remarkable smooth, with fluctuations in the height of only $\sim$30 nm over a lateral scale of 1 μm.

We can also use the force curves to determine the cornstarch Young's modulus. If we model the cantilever/particle interaction as an infinitely hard sphere (the cantilever tip) being pressed against a compliant plane (the particle surface), we can extract the

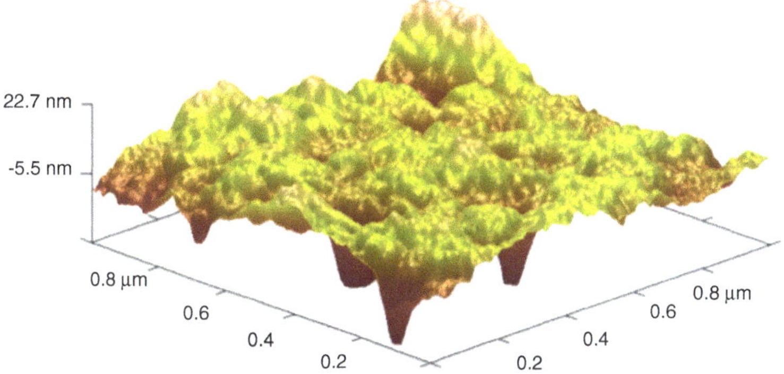

Fig. E.3 AFM 2D map of cornstarch surface height. Height profile of cornstarch particle over 1 μm^2 as determined by scanning AFM cantilever in tapping mode over particle surface. The profile reveals the particle is very smooth, with height fluctuations of only ~30 nm on a lateral scale of 1 μm

Young's Modulus by fitting the force-displacement curve to the prediction of the Derjaguin, Muller, and Toropov model [14], given by the equation

$$F_c = \frac{4}{3} E \sqrt{r_t \delta_c^3} + F_{adh}, \tag{E.1}$$

where r_t is the radius of curvature of the AFM tip and F_{adh} is the maximum adhesive force on the cantilever as it retracts from the surface. Before each set of measurements, we measured the tip radius by pressing the cantilever into a sample of known modulus, typically finding $r_r \approx 5$ nm. The adhesive force F_{adh} is the minimum of the force curve during the retraction and is determined for each tap in real time, as indicated in Fig. E.2. Finally, we note that this fitting is only performed for a particular section of retraction curve, as indicated by the bold portion of the curve in Fig. E.2. This ensures that there is mechanical contact throughout the measurement.

As with the height data, we can perform a 2D scan with the cantilever to generate plots of the modulus over large areas of the particle surface. In Fig. E.4, we show such 2D modulus maps for a dry particle and a particle that has been soaked overnight in water (examined within 1/2 h of being submerged). The maps reveal that the modulus changes slightly over the surface, ranging from an approximate lower bound of ~1.0 GPa to a maximum value of ~9.0 GPa. Additionally, we see that the variation is not noise but actually changes systematically from one region of the particle to the next. (Although this is likely not true for the particularly soft regions that appear to meander in vein-like structures, which show spatial correspondence with crevasses in the height profile.) Surprisingly, the results for the water submerged particles are not that different from the dry particles. In Fig. E.5, we plot the average modulus (defined as the average value of all the data in the 2D map) for 20 different particles (10 wet and 10 dry). These data show that dry particles are only slightly harder than wet particles, with moduli averaged over all particles of 4.9 and 3.5 GPa, respectively.

(a)
7.7 GPa
E
1.0 μm
1.7 GPa
(b)
6.8 GPa
E
1.0 μm
1.4 GPa

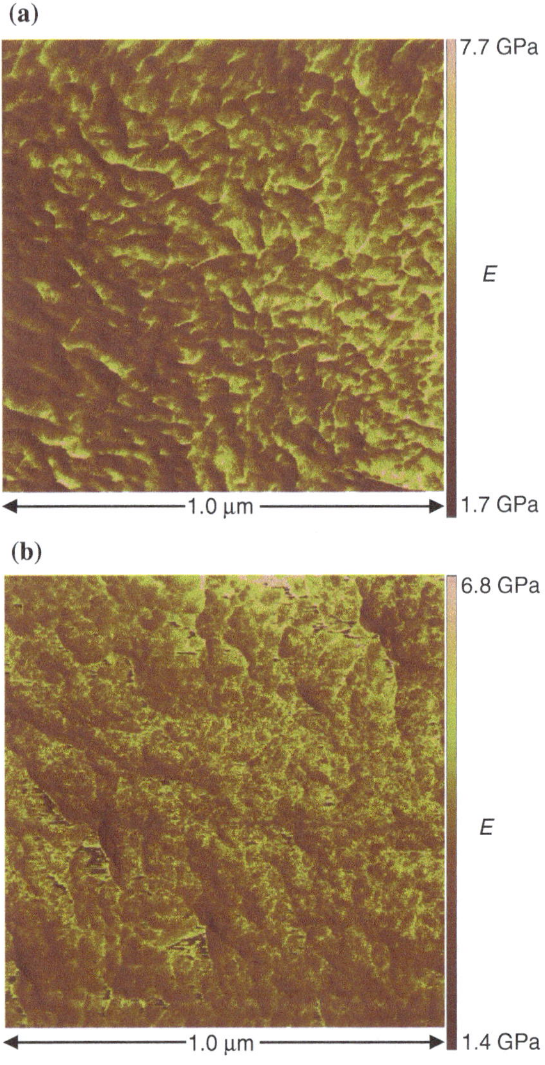

Fig. E.4 AFM 2D map of cornstarch modulus. Young's modulus E as determined from Eq. E.1 over a $1\,\mu m^2$ area of the surface for **(a)** a dry particle and **(b)** a particle ~1/2 h after being fully submerged in water for ~12 h

E (GPa)
Grain #
0
2
4
6
8
10

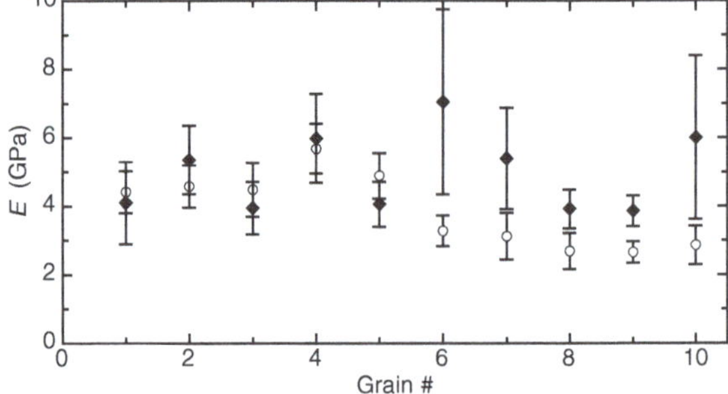

Fig. E.5 Average dry and soaked particle moduli. Spatially averaged Young's modulus E for dry particles (*solid diamonds*) and soaked particles (*open circles*) for 10 different particles. Averages over different particles are $E = 4.9\,\text{GPa}$ for the dry particles and 3.5 GPa for the wet particles

These numbers compare favorably with the results of Lionetto et al. [15] and Johnson [16], who determined the bulk modulus via speed of sound measurements to be approximately $\sim$5 and $\sim$10 GPa, respectively (note the bulk and Young's modulus should be related by a constant of order unity). For comparison, these numbers indicate that the particles are approximately as hard as ABS plastic (made to use a typical cell phone casing, for example).

Appendix F
1D Model of Particles Immersed in a Viscous Liquid

In Chap. 3, we showed that for the 2D model system of disks sitting on a plane, the relative front growth rate k was solely a function of the system packing fraction and the rake velocity. In Chap. 4, we showed that although there was speed dependence for the front growth in the cornstarch and water suspension, the increase in the saturation value as a function of packing fraction was consistent with this result (with the only difference being a different ϕ_J for the 3D system). This is interesting given that the particles in suspension are surrounded by a viscous liquid which, according to Eq. 1.2, causes their interaction to differ from the simple steric repulsion of the disk system. In this appendix, we investigate this point by computationally exploring the role lubrication forces in a simple, 1D system of particles.

We perform numerical simulations in which we dynamically calculate the position, velocity, and acceleration for a line of 1,000 particles (diameter d, initial interstitial spacing δ_0) in a liquid with dynamic viscosity η pushed from one end with a velocity v_0, as illustrated in Fig. F.1. As discussed in Chap. 2, the immersion of the particles in the liquid leads to lubrication forces between neighboring particles even before they are in contact. In the limit where the particle diameter is much larger than the interstitial spacing, the force on the ith particle is

$$F_i = \frac{3\pi\eta d^2}{8}\left(\frac{v_{i-1} - v_i}{\delta_{i-1}} + \frac{v_{i+1} - v_i}{\delta_i}\right), \tag{F.1}$$

where v_i is the velocity of the ith particle and δ_i is the spacing $x_i - x_{i-1}$ [17]. The system begins with all $v_i = 0\,\text{m/s}$ except for the leftmost particle, whose velocity is fixed at 1 m/s throughout the simulation. Subsequent values of x_i, v_i, and a_i for the rest of the particles ares found by using Newton's laws and an Eulerian time stepping algorithm with timestep $\Delta_t = 1 \times 10^{-10}\,\text{s}$. To avoid numerical issues as the spacing between particles approaches zero, we cutoff this interaction once two particles have a surface separation less than $d/1{,}000$ and force these particles to move with v_0, similar to what might be expected for particles with finite roughness.

S.R. Waitukaitis, *Impact-Activated Solidification of Cornstarch and Water Suspensions*, Springer Theses, DOI 10.1007/978-3-319-09183-9

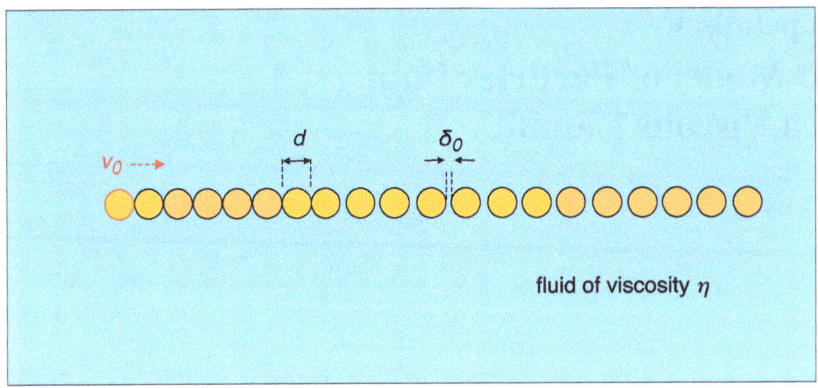

Fig. F.1 Schematic of dynamic jamming fronts with liquid-immersed particles in 1D. Liquid-immersed particles of diameter d are aligned with initial interstitial spacing δ_0 and pushed from one end with a constant velocity v_0. The dynamic viscosity of the liquid is η, leading to lubrication forces between particles before physical contact

Figure F.2a shows the velocity profiles generated from simulations with $d = 10\,\mu\text{m}$, $\delta_0 = 1\,\mu\text{m}$, $\rho = 1590\,\text{kg/m}^3$, $v_0 = 1\,\text{m/s}$, and $\eta = 1.0\,\text{cP}$ (red), 10 cP (orange), 50 cP (green) and 100 cP (blue) at two times $t_1 = 0.34\,\text{ms}$ and $t_2 = 0.68\,\text{ms}$. After a brief initialization, these profiles translate laterally at a constant velocity without changing shape. We define the solid front as the position where v_i is closest to 0.5 v_0. As can be seen from Fig. F.2b, this point is independent of η throughout the simulation. In fact, the front speed is identical to the results found in Chap. 3 (Eq. 3.9) with the initial packing fraction $\phi_0 = d/(d + \delta_0)$ and final packing fraction $\phi_J = 1.0$ (black dashed line in Fig. F.2b.

The independence of the front speed on the viscous interaction might seem surprising, especially in light of the results for shocks in granular systems above jamming [18, 19] where the speed is directly tied to the Hertzian force law between overlapping particles. In that case, however, the particle-particle interaction was largely conservative (although damping can be included, Gomez et al. showed it had little effect), whereas here the interaction is purely dissipative and strongly dependent on the relative velocity. This difference has dimensional implications. In particular, the Hertzian constant of the particles in shocks above jamming, which is proportional to the material's assumed Young's modulus E (units $\text{Kg}\,\text{m}^{-1}\,\text{s}^{-1}$), makes it possible to construct a velocity scale that is independent of the push velocity (i.e. the linear sound speed). In the present case, however, the absence of any spring-like interaction means that all velocity scales are set by v_0 and geometry.

Although the viscosity of the suspending liquid does not affect the front speed, it does affect the front width (as can be seen in Fig. F.2a). In Fig. F.2c, we zoom in on the features to show that as the viscosity is increased, so too is the front width. This effect is independent of the front broadening discussed in Chap. 3 which occurred because of the disordered nature of the system (here the "jammed" state is actually a crystal). Although we could not see the effects of this in the freely accelerating

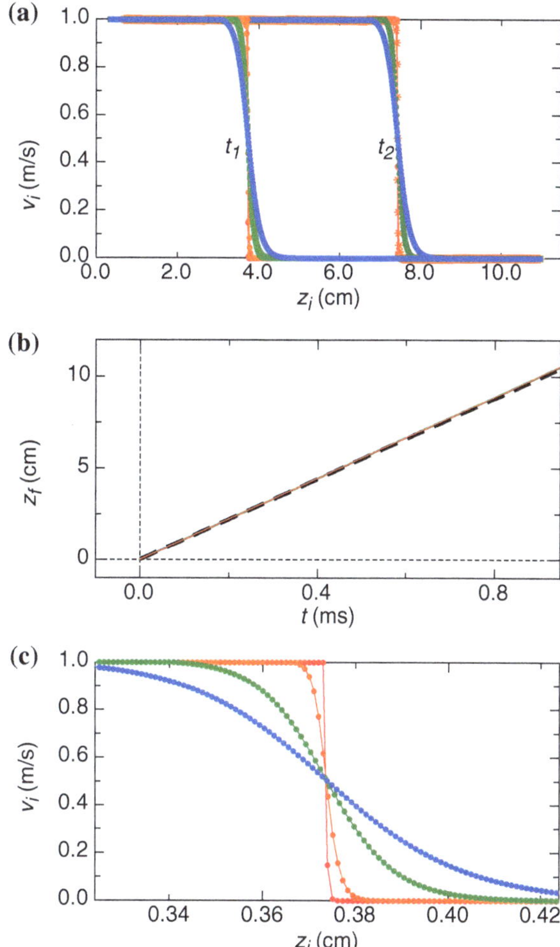

Fig. F.2 Front features with liquid-immersed particles in 1D. **a** Particle velocities v_i versus positions x_i for times 0.34 and 0.68 ms (simulation parameters $d = 10\,\mu\text{m}$, $\delta_0 = 1\,\mu\text{m}$, $\rho_{cs} = 1{,}590\,\text{kg/m}^3$, $v_0 = 1\,\text{m/s}$, and $\eta = 1.0\,\text{cP}$ (*red*), 10 cP (*orange*), 50 cP (*green*) and 10 cP (*blue*). **b** Solid front trajectory x_f versus t. Colors are the same as in (**b**). The front moves with velocity (relative to the lab frame) $x_f = v_0(d/\delta_0+1)t$ (*black dashed line*). **c** Closeup of front profiles show that front width grows with increasing viscosity

impact experiments of Chap. 2 or the speed-controlled impact experiments of Chap. 4, characterizing the relative importance of viscous broadening versus broadening brought on by disorder in liquid immersed dynamic jamming is a topic with relevance to many other phenomena and should be investigated.

References

1. M. Roche, E. Myftiu, M. C. Johnston, P. Kim, H. A. Stone, Dynamic fracture of nonglassy suspensions. Phys. Rev. Lett. **110**(14), 148304 (2013)
2. J.R. Royer, B. Conyers, E.I. Corwin, P.J. Eng, H.M. Jaeger, The role of interstitial gas in determining the impact response of granular beds. Europhys. Lett. **93**, 28008 (2011)

3. C.E. Brennen, A review of added mass and fluid inertial forces. Report to the Deptartment of the Navy (CR 82.010) (1982).
4. J.W. Glasheen, T.A. McMahon, Vertical water entry of disks at low froude numbers. Phys. Fluids. **8**, 2078–2083 (1996)
5. M. Moghisi, P.T. Squire, An experimental investigation of the initial force of impact on a sphere striking a liquid surface. J. Fluid Mech. **108**, 133–146 (1981)
6. E.A. Collins, D.J. Hoffmann, P.L. Soni, Rheology of pvc dispersions. i. effect of particle size and particle size distribution. J. Colloid. Interf. Sci. 71(1), 21–29 (1979).
7. J. Van Alsten, S. Granick, Molecular tribometry of ultrathin liquid films. Phys. Rev. Lett. **61**, 2570–2573 (1988)
8. E. Brown, H. Jaeger, Dynamic jamming point for shear thickening suspensions. Phys. Rev. Lett. **103**(8), 086001 (2009)
9. F. Boyer, É. Guazzelli, O. Pouliquen, Unifying suspension and granular rheology. Phys. Rev. Lett. **107**(18), 188301 (2011)
10. R. Seto, R. Mari, J.F. Morris, M.M. Denn, Discontinuous shear thickening of frictional hard-sphere suspensions. arXiv.org, cond-mat.soft (2013).
11. R.L. Shogren, Effect of moisture content on the melting and subsequent physical aging of cornstarch. Carbohyd. Polym. **19**(2), 83–90 (1992)
12. H. Liu, L. Yu, F. Xie, L. Chen, Gelatinization of cornstarch with different amylose/amylopectin content. Carbohyd. Polym. **65**(3), 357–363 (2006)
13. B.R. Krueger, C.A. Knutson, G.E. Inglett, C.E. Walker, A differential scanning calorimetry study on the effect of annealing on gelatinization behavior of corn starch. J. Food Sci. **52**(3), 715–718 (1987)
14. B.V. Derjaguin, V.M. Muller, Y.P. Toporov, Effect of contact deformations on the adhesion of particles. J. Colloid Interf. Sci. **53**(2), 314–326 (1975)
15. F. Lionetto, A. Maffezzoli, M.-A. Ottenhof, I.A. Farhat, J.R. Mitchell, Ultrasonic investigation of wheat starch retrogradation. J. Food Eng. **75**(2), 258–266 (2006)
16. B. Johnson, The Ultrasonic Characterization of Shear Thickening Suspensions. Ph.D. thesis, Washington University (2012).
17. L.M. Hocking, The effect of slip on the motion of a sphere close to a wall and of two adjacent spheres. J. Eng. Math. **7**(3), 207–221 (1973)
18. L. Gómez, A. Turner, M. Van Hecke, V. Vitelli, Shocks near jamming. Phys. Rev. Lett. **108**(5), 058001 (2012)
19. L. Gómez, A. Turner, V. Vitelli, Uniform shock waves in disordered granular matter. Phys. Rev. E **86**(4), 041302 (2012)

Zeitfracht Medien GmbH
Ferdinand-Jühlke-Straße 7
99095 Erfurt, Deutschland
produktsicherheit@kolibri360.de